普通高等教育“十三五”规划教材

食品质量与安全专业英语

吴 澎 张建友 杨凌宸 主编

Professional English of Food Quality and Safety

化学工业出版社
·北京·

《食品质量与安全专业英语》从食品原料安全控制、食品营养、食品添加剂、食品感官评价、食品毒理学、食品包装、食品工厂卫生、食品质量与安全管理体系、食品质量安全相关国际组织、食品质量安全认证十个方面着手编写，通过课文、词汇、课后练习、参考译文的形式概述食品质量与安全主要内容。根据学生实际需要，增加了科技论文写作专章内容，为学生学好本专业英语打下坚实基础。

《食品质量与安全专业英语》主要适用于高等院校食品质量与安全、食品科学与工程、生物工程等相关食品类本科专业的教学用书，也可用于相关专业研究生、专科生的辅助教学参考用书；同时，也可作为食品从业人员掌握食品安全中、英文双语词汇的辅助教材。

图书在版编目（CIP）数据

食品质量与安全专业英语/吴澎，张建友，杨凌宸主编.
—北京：化学工业出版社，2017.9（2025.3 重印）
普通高等教育“十三五”规划教材
ISBN 978-7-122-30475-9

Ⅰ.①食…　Ⅱ.①吴…②张…③杨…　Ⅲ.①食品安全-质量管理-英语-高等学校-教材　Ⅳ.①TS201.6

中国版本图书馆 CIP 数据核字（2017）第 201194 号

责任编辑：尤彩霞　　文字编辑：向　东
责任校对：宋　夏　　装帧设计：史利平

出版发行：化学工业出版社（北京市东城区青年湖南街 13 号　邮政编码 100011）
印　　装：北京虎彩文化传播有限公司
787mm×1092mm　1/16　印张 9¼　字数 233 千字　2025 年 3 月北京第 1 版第 6 次印刷

购书咨询：010-64518888　　售后服务：010-64518899
网　　址：http://www.cip.com.cn
凡购买本书，如有缺损质量问题，本社销售中心负责调换。

定　　价：35.00 元

《食品质量与安全专业英语》编者名单

主　　编：吴　澎　张建友　杨凌宸

副 主 编：路　飞　王　颖　王晶晶　张一敏

参编人员（姓名汉语拼音排序）：

Carol Yongmei Zhang　英国皇家农业大学

冯　颖　沈阳农业大学

郭　瑜　山西农业大学

李　可　郑州轻工业学院

李燮昕　四川旅游学院

路　飞　沈阳师范大学

齐　丽　泰安市委党校

任海伟　兰州理工大学

孙炳新　沈阳农业大学

唐彦君　黑龙江八一农垦大学

王晶晶　锦州医科大学

王　颖　黑龙江八一农垦大学

吴　澎　山东农业大学

杨凌宸　湖南农业大学

于立梅　仲恺农业工程学院

袁　媛　吉林大学

张建友　浙江工业大学

张一敏　山东农业大学

近些年来，食品安全引发的社会、政治和贸易问题逐渐增多，世界各国的食品安全管理体系也随之不断完善。 及时了解和掌握各国在食品安全管理方面的相关研究成果以及适合中国国情的食品安全管理体系，是食品质量与安全专业课程建设亟须解决的问题。 本教材紧密结合我国急需食品安全管理人才的实际情况，填补我国食品质量与安全专业多年来未有及时更新的专业英语教材的空白，适应当前各高校食品课程开设需要。 编者近年已出版多本教材，本书联系各高校共同编写，针对本科生、研究生急需教材的情况进行整理编写，以期希望能够解决教学现实问题。

《食品质量与安全专业英语》教材从食品原料安全控制、食品营养、食品添加剂、食品感官评价、食品毒理学、食品包装、食品工厂卫生、食品质量与安全管理体系、食品质量安全相关国际组织、食品质量安全认证十个方面着手，通过课文、词汇、课后练习、参考译文的形式概述食品质量与安全主要内容。 根据学生实际需要，增加了科技论文写作专章，为学生学好专业英语打下坚实基础。

由于时间仓促，编者水平有限，书中不当之处在所难免，敬请读者批评指正，以便我们修订时改正。

编者

2017 年 9 月

目录
Contents

Unit 1

Food Materials Safety Control

1.1 Food Safety Hazard

A food-borne hazard is a biological, chemical, or physical agent in, or condition of food with the potential to cause an adverse health effect.

1.1.1 Biological Hazards

Biological hazards include pathogenic bacteria, fungi, viruses, prions, protozoans and helminthic parasites. There are two types of foodborne disease from microbial pathogens: infections and intoxications. Infections result from ingestion of live pathogenic organisms that multiply within the body and produce disease. Intoxications occur when toxins produced by pathogens are consumed. Intoxications can occur even if no viable microorganisms are ingested. This often occurs when foods are stored under conditions that allow the pathogens to grow and produce toxin. Subsequent processing of the food may destroy the microorganisms but not the toxin. Manifestations of biological hazards typically involve foodborne illnesses with symptoms including gastrointestinal distress, diarrhoea, vomiting and sometimes death. The U. S. Public Health Service classifies moist, high-protein, and low-acid foods as potentially hazardous. High-protein foods consist, in whole or in part, of milk or milk products, shell eggs, meats, poultry, fish, shellfish, edible crustacea (shrimp, lobster, crab). Baked or boiled potatoes, tofu and other soy protein foods, plant foods that have been heat-treated, and raw seed sprouts (such as alfalfa or bean sprouts) also pose a hazard. These foods can support rapid growth of infectious or disease-causing microorganisms.

1.1.2 Chemical Hazards

A very wide variety of chemical hazards may appear in food products either by natural occurrence in a raw material or by deliberate or unintentional addition during processing. When ingested, these may cause gastrointestinal distress, organ damage and immunological reactions that may result in death. The long-term ingestion of foods containing toxic chemicals can lead to chronic effects, including cancer. Some chemical hazards are presented below:

Naturally occuring chemical hazards: allergens, mycotoxins.

Agricultural chemicals: pesticides, fertilizers, antibiotics.

Heavy metals: lead, cadmium, mercury, arsenic, uranium.

Chemicals used in food processing environments: lubricants, detergents, sanitizers, refrigerants, pesticides.

Chemicals used in food packaging materials: plasticizers, other additives used in manufacture.

1.1.3 Physical Hazards

Food-borne physical hazards are commonly called "foreign materials" or "foreign bodies". The general sources of contamination are the environment, the food itself, the food processing facilities and personal objects. Their presence in food may result in choking, or oralor internal cuts, but rarely result in death.

Soil and stones are typical environmental contaminants during harvesting. Some foreigh material contamination originates with the food itself such as fruit stones and stems, bones or bone chips from fish and meat. A great deal of foreigh material contamination originates in food processing facilities. Nails, cut wires and broken utility blades can be dropped into the food stream by maintenance workers. Pieces of glass, hard plastic and wood splinters can enter the food from other fixtures and utensils in the processing area. Personal objects used or worn by maintenance and line workers and by food handlers in food service operations often fall into the food. These may include rings, pencils, papers, earrings, nose rings, buttons, thermometers, hair and gloves.

1.2 Control of Microbiological Hazards

Some educators refer to the "three Ks" as a comprehensive programme of microbiological hazard control in food production, those being, "kill them", "keep them from growing" and "keep them out". Described more scientifically, these three procedures are:

1.2.1 Destruction of Microorganisms

Many well-established and several novel procedures are available to kill microorganisms. These include thermal processes such as pasteurisation and ultra-high temperatures sterilisation, and non-thermal processes such as irradiation, high hydrostatic pressure and pulsed electric fields. Most often, pasteurisation involves a cooking or heating procedure conducted at atmospheric pressure. It is used to protect the public health by killing pathogenic microorganisms and to extend product shelf-life by killing spoilage microorganisms. Many foods to be sterilised are packaged into metal, glass or plastic retail containers, hermetically sealed and processed under pressure with steam at 121℃ or higher. Foods may be sterilised at ultra-high temperatures (UHT) for a very short time, e. g. 140～150℃ for several seconds.

1.2.2 Prevention of Microbial Growth

The principal process controls to prevent the growth of microorganisms in foods include refrigeration, freezing, modified atmospheres. Below its minimum growth temperature, a microorganism is no longer able to grow. The generally accepted temperature for optimum refrigeration is 5℃ or lower. The shelf-life of refrigerated perishable foods can often be extended by frozen storage. Commercially produced frozen foods are usually stored at －18℃. Some perishable food products are packaged in containers with a headspace for air under atmospheric pressure. Other factors permitting, aerobic microorganisms can grow freely in such products. Their growth can be inhibited or prevented by the removal of headspace oxygen (vacuum packaging) or the addition of inhibitory gases (modified atmosphere packaging).

1.2.3 Prevention of Contamination

Many potential microbiological hazards can be avoided by preventing cross-contamination. Cross-contamination is the transportation of harmful substances to food by:

Hands that touch raw foods, such as chicken, then touch food that will not be cooked, like salad ingredients.

Surfaces, like cutting boards or cleaning cloths, that touch raw foods, are not cleaned and sanitized, then touch ready-to-eat food.

Raw or contaminated foods are touch or drip fluids on cooked or ready-to-eat foods.

Cleaning and sanitation procedures and personnel practices used in food processing facilities are most important in this regard.

1.3 Control of Chemical Hazards

Food manufacturing facilities should maintain a chemical control plan to prevent contamination of its products with allergens, mislabelled or adulterated ingredients, and cleaning and maintenance chemicals. As with all food safety and quality practices, employee training and awareness is an essential factor in minimising the risk of chemical hazards in foods. For example, these include verification of the accuracy of the ingredient declarations on product labels, and the implementation of suitable prerequisite control for the receipt, storage and use of high-risk ingredients. Quality assurance and production personnel must verify that the ingredients are added correctly to product mixers or preblend operation. Cleaning and sanitation chemicals, pesticides and other non-food chemicals need to be stored in a confined, locked area and not to be used during periods of food production.

1.4 Control of Physical Hazards

There are three principal means to control physical hazards in foods:

① Exclusion—including programmes for glass, wood, personnel practices and pest control

For example, most food processing facilities maintain a strict prohibition on the use of glass or brittle plastic instruments, utensils to avoid the possible entry of glass fragments into the food product. Wooden pallets and wooden handles on tools and maintenance equipment used in all production areas need to be excluded. Employees should not wear items of jewellery and should wear uniforms and hair covers. The uniforms should have no pockets, so that items such as pens and pencils cannot be carried in the pockets and fall into the products. Insect activity can be minimised inside facilities by the use of insect light traps that attract flying insects with an ultraviolet light.

② Removal—by the use of devices such as magnets, sifters

Many types of in-line magnets are used by food processors on incoming ingredients, processing equipment and packaging operations, both to protect the equipment from damage by tramp metal and to avoid product contamination. Sifters and sieves are often used to separate foreign materials from dry food materials.

③ Detection—by using instruments such as metal detectors, X-rays and optical sorters

Several technologies are available to detect the presence of foreign materials in a food. The most common of these are metal detectors that can be used online or for packaged products. With image-enhancing capabilities, X-ray devices can be used to detect dense foreign material inside food products, such as bone chips in meat products. Optical technologies, using visible or ultraviolet light, are used with fruits, vegetables and nuts to detect surface defects and the presence of extraneous vegetable matter, stones, etc.

Technical Terms

food-borne disease 食源性疾病
pathogenic bacteria 病原性细菌
fungi ['fʌŋgiː] *n*. 真菌
virus ['vaɪrəs] *n*. 病毒
prion ['priːɒn] *n*. 感染性蛋白质
protozoan [ˌprəʊtə'zəʊən] *n*. 原生动物
helminthic parasite 蠕虫寄生虫
intoxication [ɪnˌtɒksɪ'keɪʃn] *n*. 中毒
gastrointestinal [ˌgæstrəʊɪn'testɪnl] *adj*. 胃与肠的
diarrhoea [ˌdaɪə'rɪə] *n*. 腹泻
allergen ['ælədʒən] *n*. 过敏源
mycotoxin [ˌmaɪkəʊ'tɒksən] *n*. 毒枝菌素，真菌毒素
alkaloid ['ælkəlɔɪd] *n*. 生物碱
phytohaemagglutinin [faɪtəhiːmæg'luːtɪnɪn] *n*. 植物血凝素
pesticide ['pestɪsaɪd] *n*. 杀虫剂，农药
fertilizer ['fɜːtəlaɪzə(r)] *n*. 肥料，化肥
antibiotic [ˌæntibaɪ'ɒtɪk] *n*. 抗生素，抗菌素
lead [liːd] *n*. 铅
cadmium ['kædmiəm] *n*. 镉
mercury ['mɜːkjəri] *n*. 汞，水银
arsenic ['ɑːsnɪk] *n*. 砷，砒霜
uranium [ju'reɪniəm] *n*. 铀
lubricant ['luːbrɪkənt] *n*. 润滑剂，润滑油
refrigerant [rɪ'frɪdʒərənt] *n*. 制冷剂，冷冻剂
plasticizer ['plæstɪsaɪzə] *n*. 增塑剂，可塑剂
vinyl chloride 氯乙烯
adhesive [əd'hiːsɪv] *n*. 黏合剂
tin [tɪn] *n*. 锡，马口铁
thermal processes 热处理
pasteurisation [ˌpɑːstəraɪ'zeɪʃən] *n*. 巴斯德杀菌法
sterilisation [ˌsterɪlaɪ'zeɪʃən] *n*. 消毒，灭菌

non-thermal processes 非热加工
high hydrostatic pressure 高静压技术
pulsed electric fields 脉冲电场
ultra-high temperatures (UHT) 超高温
refrigeration [rɪˌfrɪdʒə'reɪʃn] *n.* 冷藏
freezing ['friːzɪŋ] *n.* 冻结，结冰
frozen storage 冻藏
vacuum packaging 真空包装
modified atmosphere packaging 气调包装
ready-to-eat food 方便食品，即食食品
cross-contamination 交叉污染
sanitize ['sænɪtaɪz] *v.* 进行消毒
cleaning and sanitation procedures 清洗消毒步骤
adulterate [ə'dʌltəreɪt] *v.* 掺假
sifter ['sɪftə(r)] *n.* 筛具

Exercises

Ⅰ. Answer the following questions according to the article

1. What is the definition of a foodborne hazard?
2. What do biological hazards mainly include?
3. What disease systems can be caused by chemical hazards?
4. How to control microbiological hazards?
5. List some measures that can be taken to control physical hazards.

Ⅱ. Choose a term from what we have learnt to fill in each of the following blanks. Change the word form where necessary

1. Foodborne hazards can be individed into __________, __________ and __________ three categories.

2. __________, __________ and __________ foods are classified as potentially biological hazardous because these foods can support rapid growth of infectious or disease-causing microorganisms.

3. Microorganisms in food materials can be killed by thermal processes such as __________, and non-thermal processes such as __________.

4. __________ are naturally occuring protein to which some person develop a hypersensitivity or immunological response.

5. Peanut product producers use __________ to remove dark-coloured nuts, thereby reducing the risk of aflatoxin contamination in finished products.

Unit 2

Food Nutrition

Food and food products have become commodities produced and traded in a market that has expanded from an essentially local base to an increasingly global one. Changes in the world food economy are reflected in shifting dietary patterns, for example, increased consumption of energy-dense diets high in fat, particularly saturated fat, and low in unrefined carbohydrates. These patterns are combined with a decline in energy expenditure that is associated with a sedentary lifestyle—motorized transport, labour-saving devices in the home, the phasing out of physically demanding manual tasks in the workplace, and leisure time that is preponderantly devoted to physically undemanding pastimes.

Furthermore, rapid changes in diets and lifestyles that have occurred with industrialization, urbanization, economic development and market globalization have accelerated over the past decade. This is having a significant impact on the health and nutritional status of populations, particularly in developing countries and in countries in transition. Nutrition is coming to the fore as a major modifiable determinant of chronic disease, with scientific evidence increasingly supporting the view that alterations in diet have strong effects, both positive and negative, on health throughout life.

2.1 Food and Nutrition

Food refers to what humans eat for the consumption of more than one nutrient, which is composed of carbohydrates, fats, proteins and water. Almost all foods are of plant or animal origin expect some other sources such as fungi. Fungi is used in the preparation of fermented and pickled foods. When food is abundant, it is wasted or treated as a commodity. However, when food is scarce, it is regarded as the staff of life and its distribution becomes a highly emotional issue. Food production worldwide is increasing faster than population, but distribution is uneven, reserves are limited, and bad weather conditions could lead to widespread famines.

Nutrition refers to both the absorption and usage of food, including all the elements that are intaked, digested, absorbed in the human body. An extremely important aspect of food is nutrition, which is especially true for vulnerable groups such as infants and young children. Therefore, we use accurate diet assessment which is important for assessing the relationship of diet to health outcomes; identifying factors (e. g. biological, environmental, psychosocial) influencing dietary intake which could be targeted for change; assessing changes over time in diet (surveillance) and evaluating the outcome of dietary interventions. Yet accurate dietary assessment remains a challenge; perhaps especially when children are the subjects.

2. 1. 1 The Nutrients

Nutrients refers to substances which have trophic function, including proteins, lipids, carbohydrates, vitamins, minerals and water. All of the nutrients except for mineral elements and water are calssified as organic chemicals because they contain carbon.

About 60% of the human body is water by weight, which is constantly lost from the body and must constantly be replaced. Carbohydrates and fats or lipids are especially important energy-yielding nutrients. Proteins play a big role in functionality such as texture, water binding, emulsification, gel formation, and foam formation.

Vitamins are organic chemicals, other than essential amino acids and fatty acids that must be supplied to an animal in small amounts to maintain health. The vitamins are divided into two general categories based on their solubility in either water or fat. The fat-soluble vitamins are vitamins A, D, E, K; the water-soluble vitamins include vitamin C (ascorbic acid), niacin, thiamin, riboflavin, folacin (folic acid), pantothenic acid, pyridoxine, vitamin B_{12} and biotin.

Depending on whether or not they are required for human nutrition and have metabolic roles in the body, the mineral elements are classified as either essential or non-essential. Some examples of essential ones are sodium, potassium, calcium, and phosphorus. Some examples of non-essential ones are iron, iodine, manganese, zinc, and fluorine.

2. 1. 2 Nutrients and Food Processing

Food processing is the methods and techniques used to transform raw ingredients into food for human consumption. The effects on the losses of vitamins and minerals of cereals cooked by different methods are different. The losses of minerals are mainly water-soluble losses. The losses of B vitamins in cooking include water-soluble loss, heating loss, oxidation loss and alkali treatment loss. Vitamin B_1 often reaches the highest loss rate of vitamins .

In the processing of tofu products, the mineral content is often increased. Soybean itself is rich in calcium, Tofu is an important source of dietary calcium and magnesium elements because calcium or magnesium salt is used as tofu coagulant. The micronutrients in soybean are basically preserved in soybean products. However, there are still some B vitamins lost in water-soluble loss. Most of them are lost in precipitation and some loss is caused by heating degradation.

Fresh vegetables is the main consumption form in daily diet. There will be significantly changes in vitamins and minerals contents in vegetables because of the processing of curing,

drying, quick freezing and canning. In particular, the water-soluble loss and decomposition loss of vitamin C, vitamin B_1 and folic acid will caused by heating.

In the processing of meat, poultry and fish, the loss of water-soluble vitamins is more obviously than the losses of proteins and minerals. The fat content will be changed according to different processing methods.

Dairy products are a kind of food that is rich in nutritions. There will be little effects on the milk proteins but different levels of losses of vitamins and minerals in the proper processing methods.

2.2 Nutrition Education

The increase in obesity and chronic diseases such as diabetes and heart disease worldwide reflects the complex interactions of biology, personal behaviour and environment. Consequently, there has been a greater recognition of the importance of nutrition education. Nutrition education needs to address food preferences and sensory-affective factors; person-related factors such as perceptions, beliefs, attitudes, meanings, and social norms; and environmental factors.

Nutrition education has been defined as "any combination of educational strategies, accompanied by environmental supports, designed to facilitate voluntary adoption of food choices and other food and nutrition-related behaviors conducive to health and well-being; nutrition education is delivered through multiple venues and involves activities at the individual, community, and policy levels."

Technical Terms

scarce [skeəs] *adj*. 缺乏的，不足的；稀有的
commodity [kə'mɒdəti] *n*. 商品，货物；日用品
intake ['ɪnteɪk] *n*. 摄入量
digest [daɪ'dʒest] *vt*. 消化，吸收
absorb [əb'sɔːb] *vt*. 吸收，吸引
diet assessment 膳食评估
fungi ['fʌŋgiː] *n*. 真菌，菌类；蘑菇
fermented [fə'mentid] *adj*. 酿造，已发酵的
pickled ['pɪkld] *adj*. 腌制的，盐渍的；烂醉如泥的
carbohydrates [ˌkɑːbəʊ'haɪdreɪt] *n*. 糖类；碳水化合物
lipids ['lɪpɪdz] *n*. 脂肪，油脂；脂类
proteins [p'rəʊtiːnz] *n*. 蛋白质
riboflavin [ˌraɪbəʊ'fleɪvɪn] *n*. 核黄素，维生素 B_2
folacin ['fɒləsɪn] *n*. 叶酸
pantothenic acid [ˌpæntə'θenik'æsid] *n*. 泛酸
pyridoxine [pɪrɪ'dɒksiːn] *n*. 维生素 B_6，吡哆醇
biotin ['baɪətɪn] *n*. 维生素 H，生物素
sodium ['səʊdiəm] *n*. 钠
calcium ['kælsiəm] *n*. 钙
iron ['aɪən] *n*. 铁
nutritional status 营养状况
nutritive value 营养价值

food processing 食品加工

Exercises

Ⅰ. Answer the following questions according to the article

1. What is the definition of nutrition?
2. How many categories can nutrition be divided into?
3. What is the definition of food?
4. Name five minerals required by the body.
5. How many categories can mineral be divided into?

Ⅱ. Judge whether the sentence is right or wrong

1. People who are sleeping consume only ree. (　　)
2. People wish to lose weight can do no better than to control calories and avoid carbohydrate-rich foods. (　　)
3. Starch is a kind of carbohydrates. (　　)
4. Glycogen can be hydrolyzed into glucose and fructose. (　　)
5. The calories that burned in the involuntary work remain constant during human life. (　　)

Unit 3

Food Additives

For centuries, ingredients have served useful functions in a variety of foods. Our ancestors used salt to preserve meats and fish, added herbs and spices to improve the flavor of foods, preserved fruit with sugar, and pickled cucumbers in a vinegar solution. Today, consumers demand and enjoy a food supply that is flavorful, nutritious, safe, convenient, colorful and affordable. Food additives and advances in technology help make that possible.

There are thousands of food additives used to make foods. For example, The government of the People's Republic of China maintains a list of over 2000 additives in its data base while over 3000 in the United States', many of which we use at home every day (e. g., sugar, baking soda, salt, vanilla, yeast, spices and colors).

Still, some consumers have concerns about additives because they may see the long, unfamiliar names and think of them as complex chemical compounds. In fact, every food we eat—whether a just-picked strawberry or a homemade cookie—is made up of chemical compounds that determine flavor, color, texture and nutrient value. All food additives are carefully regulated by governments and various international organizations to ensure that foods are safe to eat and are accurately labeled.

3.1 Concept of Food Additives

In its broadest sense, a food additive is any substance added to food. Legally, the term refers to "any substance the intended use of which results or may reasonably be expected to result—directly or indirectly—in its becoming a component or otherwise affecting the characteristics of any food." This definition includes any substance used in the production, processing, treatment, packaging, transportation or storage of food. The purpose of the legal definition, however, is to impose a premarket approval requirement. Therefore, this definition excludes ingredients whose use is generally recognized as safe (where government approval is not needed); those ingredients approved for use by a government prior to the food additives provisions of law, and color additives and pesticides where other legal premarket approval requirements apply.

Direct food additives are those that are added to a food for a specific purpose in that food. For example, xanthan gum—used in salad dressings, chocolate milk, bakery fillings, puddings and other foods to add texture—is a direct additive. Most direct additives are identified on the ingredient label of foods.

Indirect food additives are those that become part of the food in trace amounts due to its packaging, storage or other handling. For instance, minute amounts of packaging substances may find their way into foods during storage. Food packaging manufacturers must prove to the government that all materials coming in contact with food are safe before they are permitted for use in such a manner.

3.2 Functions of Food Additives

Additives perform a variety of useful functions in foods that consumers often take for granted. Some additives could be eliminated if we were willing to grow our own food, harvest and grind it, spend many hours cooking and canning, or accept increased risks of food spoilage. However, most consumers today rely on the many technological, aesthetic and convenient benefits that additives provide.

The following are main reasons why additives are added to foods:

(1) To Maintain or Improve Safety and Freshness

Preservatives slow product spoilage caused by mold, air, bacteria, fungi or yeast. In addition to maintaining the quality of the food, they help control contamination that can cause foodborne illness, including life-threatening botulism. One group of preservatives—antioxidants—prevents fats, oils, and the foods containing them from becoming rancid or developing an off-flavor. They also prevent cut fresh fruits such as apples from turning brown when exposed to air.

(2) To Improve or Maintain Nutritional Value

Vitamins and minerals (and fiber) are added to many foods to make up for those lacking in a person's diet or lost in processing, or to enhance the nutritional quality of a food. Such fortification and enrichment has helped reduce malnutrition worldwide. All products containing added nutrients must be appropriately labeled.

(3) To Improve Taste, Texture and Appearance

Spices, natural and artificial flavors, and sweeteners are added to enhance the taste of food. Food colors maintain or improve appearance. Emulsifiers, stabilizers and thickeners give foods the texture and consistency consumers expect. Leavening agents allow baked goods to rise during baking. Some additives help control the acidity and alkalinity of foods, while other ingredients help maintain the taste and appeal of foods with reduced fat content.

3.3 Types of Food Additives

Food additives can be divided into six major categories: preservatives, nutritional additives, coloring agents, flavoring agents, texturizing agents, and miscellaneous additives. A detailed list of these additives are noted in Table 1.

(1) Preservatives

There are basically three types of preservatives used in foods: antimicrobials, antioxidants, and antibrowning agents.

Antimicrobials play a major role in extending the shelf-life of numerous snack and convenience foods and have come into even greater use in recent years as microbial food safety concerns have increased.

The antioxidants are used to prevent lipid and/or vitamin oxidation in food products. They are used primarily to prevent autoxidation and subsequent development of rancidity and off-flavor. They vary from natural substances such as vitamins C and E to synthetic chemicals such as butylated hydroxyanisole (BHA) and butylated hydroxytoluene (BHT).

The antioxidants are especially useful in preserving dry and frozen foods for an extended period of time. Antibrowning agents are chemicals used to prevent both enzymatic and non-enzymatic browning in food products, especially dried fruits or vegetables.

(2) Nutritional Additives

Nutritional additives have increased in use in recent years as consumers have become more concerned about and interested in nutrition. Nutritional additives includes mainly vitamins and minerals.

Vitamins, which are also used in some cases as preservatives, are commonly added to cereals and cereal products to restore nutrients lost in processing or to enhance the overall nutritive value of the food. The addition of vitamin D to milk and of B vitamins to bread has been associated with the prevention of major nutritional deficiencies.

Minerals such as iron and iodine have also been of extreme value in preventing nutritional deficiencies. Like vitamins, the primary use of minerals is in cereal products.

(3) Coloring Agents

A color additive is any dye, pigment or substance which when added or applied to a food, drug or cosmetic, or to the human body, is capable (alone or through reactions with other substances) of imparting color. In addition, sodium nitrite is used not only as an antimicrobial, but also to fix the color of meat by interaction with meat pigments. Our government is responsible for regulating all color additives to ensure that foods containing color additives are safe to eat, containing only approved ingredients and are accurately labeled.

Color additives are used in foods for many reasons: ①to offset color loss due to exposure to light, air, temperature extremes, moisture and storage conditions; ②to correct natural variations in color; ③to enhance colors that occur naturally; and ④to provide color to colorless and "fun" foods. Without color additives, colas would not be brown, margarine would not be yellow and mint ice cream would not be green. Color additives are now recognized as an important part of practically all processed foods we eat.

In America the Government's permitted colors are classified as subject to certification or exempt from certification, both of which are subject to rigorous safety standards prior to their approval and listing for use in foods.

Certified colors are synthetically produced (or human made) and used widely because they impart an intense, uniform color, are less expensive, and blend more easily to create a variety of hues. There are nine certified color additives approved for use at present. Certified food colors generally do not add undesirable flavors to foods.

Colors that are exempt from certification include pigments derived from natural sources such as vegetables, minerals or animals. Nature derived color additives are typically more expensive than certified colors and may add unintended flavors to foods. Examples of exempt colors include annatto extract (yellow), dehydrated beets (bluish-red to brown), caramel (yellow to tan), beta-carotene (yellow to orange) and grape skin extract (red, green).

Certified color additives are categorized as either dyes or lakes. Dyes dissolve in water are manufactured as powders, granules, liquids or other special-purpose forms. They can be used in beverages, dry mixes, baked goods, confections, dairy products, pet foods and a variety of other products.

Lakes are the water insoluble form of the dye. Lakes are more stable than dyes and are ideal for coloring products containing fats and oils or items lacking sufficient moisture to dissolve dyes. Typical uses include coated tablets, cake and donut mixes, hard candies and chewing gums.

(4) Flavoring Agents

Flavoring agents comprise the greatest number of additives used in foods. There are three major types of flavoring additives: sweeteners, natural and synthetic flavors, and flavor enhancers.

The most commonly used sweeteners are sucrose, glucose, fructose, and lactose, with sucrose being the most popular. These substances, however, are commonly classified as foods rather than as additives. The most common additives used as sweeteners are lowcalorie or noncaloric sweeteners such as saccharin and aspartame.

In addition to sweeteners, there are more than 1700 natural and synthetic substances used to flavor foods. These additives are, in most cases, mixtures of several chemicals and are used to substitute for natural flavors. In most cases, flavoring agents are the same chemical mixtures that would naturally provide the flavor.

Flavor enhancers magnify or modify the flavor of foods and do not contribute any flavor of their own. Flavor enhancers, which include chemicals such as monosodium glutamate, disodium 5′-inosinate and disodium 5′-guanylate, are often used in daily foods or in soups to enhance the perception of other tastes.

(5) Texturizing Agents

Although flavoring agents comprise the greatest number of chemicals, texturizing agents are used in the greatest total quantity. These agents are used to add to or modify the overall texture or mouthfeel of food products. Emulsifiers and stabilizers are the primary additives in this category.

Emulsifiers include natural substances such as lecithin and mono-and diglycerides as well as several synthetic derivatives. The primary role of these agents is to allow flavors and oils to be dispersed throughout a food product.

Stabilizers include several natural gums such as carrageenan as well as natural and modified starches. These additives have been used for several years to provide the desired texture in products such as ice cream and are now also finding use in both dry and liquid products. They also are used to prevent evaporation and deterioration of volatile flavor oils.

Humectants, phosphates, are often used to modify the texture of foods containing protein or starch. These chemicals are especially useful in stabilizing various dairy and meat products. The phosphates apparently react with protein and/or starch and modify the water-holding capacity of these natural food components.

Dough conditioners, such as steroyl-2-lactylate, L-cysteine, have many advantages attributed to dough conditioners include ①improved tolerance to variations in flour and other ingredient quality, ②doughs with greater resistance to mixing and mechanical abuse, ③better gas retention resulting in lower yeast requirements, shorter proof times, and increased baked product volume, ④increased uniformity in cell size, a finer grain, and a more resilient texture, and ⑤improved slicing. They are also used as texturizing agents under very specific conditions.

(6) Miscellaneous Additives

There are numerous other chemicals used in food products for specific yet limited purposes. Included are various processing aids such as chelating agents, enzymes, and antifoaming agents; surface finishing agents; catalysts; and various solvents, lubricants, and propellants.

The following summary lists the types of common food additives, names, function and examples. Some additives are used for more than one purpose.

Table 1. List of food additives

Types	Names	Function	Examples
Preservatives	Ascorbic acid, citric acid, sodium benzoate, calcium propionate, sodium erythorbate, sodium nitrite, calcium sorbate, potassium sorbate, BHA, BHT, EDTA, tocopherols (vitamin E)	Prevent food spoilage from bacteria, molds, fungi, or yeast (antimicrobials); slow or prevent changes in color, flavor, or texture and delay rancidity(antioxidants); maintain freshness	Fruit sauces and jellies, beverages, baked goods, cured meats, oils and margarines, cereals, dressings, snack foods, fruits and vegetables
Sweeteners	Sucrose (sugar), glucose, fructose, sorbitol, mannitol, corn syrup, high fructose corn syrup, saccharin, aspartame, sucralose, acesulfame potassium (acesulfame-K), neotame	Add sweetness with or without the extra calories	Beverages, baked goods, confections, table-top sugar, substitutes, many processed foods
Color Additives	FD&C Blue Nos. 1 and 2, FD&C Green No.3, FD&C Red Nos. 3 and 40, FD&C Yellow Nos.5 and 6, Orange B, Citrus Red No. 2, annatto extract, beta-carotene, grape skin extract, cochineal extract or carmine, paprika oleoresin, caramel color, fruit and vegetable juices, saffron (Note: Exempt color additives are not required to be declared by name on labels but may be declared simply as colorings or color added)	Offset color loss due to exposure to light, air, temperature extremes, moisture and storage conditions; correct natural variations in color; enhance colors that occur naturally; provide color to colorless and "fun" foods	Many processed foods, (candies, snack foods margarine, cheese, soft drinks, jams/jellies, gelatins, pudding and pie fillings)
Flavors and Spices	Natural flavoring, artificial flavor, and spices	Add specific flavors (natural and synthetic)	Pudding and pie fillings, gelatin dessert mixes, cake mixes, salad dressings, candies, soft drinks, ice cream, BBQ sauce

续表

Types	Names	Function	Examples
Flavor Enhancers	Monosodium glutamate (MSG), hydrolyzed soy protein, autolyzed yeast extract, disodium guanylate or inosinate	Enhance flavors already present in foods (without providing their own separate flavor)	Many processed foods
Nutrients	Thiamine hydrochloride, riboflavin (vitamin B_2), niacin, niacinamide, folate or folic acid, beta carotene, potassium iodide, iron or ferrous sulfate, alpha tocopherols, ascorbic acid, vitamin D, amino acids (L-tryptophan, L-lysine, L-leucine, L-methionine)	Replace vitamins and minerals lost in processing (enrichment), add nutrients that may be lacking in the diet (fortification)	Flour, breads, cereals, rice, macaroni, margarine, salt, milk, fruit beverages, instant breakfast drinks
Emulsifiers	Soy lecithin, mono-and diglycerides, egg yolks, polysorbates, sorbitan monostearate	Allow smooth mixing of ingredients, prevent separation, keep emulsified products stable, reduce stickiness, control crystallization, keep ingredients dispersed, and to help products dissolve more easily	Salad dressings, peanut butter, chocolate, margarine, frozen desserts
Stabilizers and Thickeners, Binders, Texturizers	Gelatin, pectin, guar gum, carrageenan, xanthan gum, whey	Produce uniform texture, improve "mouth-feel"	Frozen desserts, dairy products, cakes, pudding and gelatin mixes, dressings, jams and jellies, sauces
pH Control Agents and Acidulants	Lactic acid, citric acid, ammonium hydroxide, sodium carbonate	Control acidity and alkalinity, prevent spoilage	Beverages, frozen desserts, chocolate, low acid canned foods, baking powder
Leavening Agents	Baking soda, monocalcium phosphate, calcium carbonate	Promote rising of baked goods	Breads and other baked goods
Anti-caking Agents	Calcium silicate, iron ammonium citrate, silicon dioxide	Keep powdered foods free-flowing, prevent moisture absorption	Salt, baking powder, confectioner's sugar
Humectants	Glycerin, sorbitol	Retain moisture	Shredded coconut, marshmallows, soft candies, confections
Yeast Nutrients	Calcium sulfate, ammonium phosphate	Promote growth of yeast	Breads and other baked goods
Dough Strengtheners and Conditioners	Ammonium sulfate, azodicarbonamide, L-cysteine	Produce more stable dough	Breads and other baked goods
Firming Agents	Calcium chloride, calcium lactate	Maintain crispness and firmness	Processed fruits and vegetables, soybean products
Enzyme Preparations	Lactase, papain, rennet, chymosin	Modify proteins, polysaccharides and fats	Cheese, dairy products, meat

3.4 Additives Approved for Use in Foods

Today, food additives are more strictly studied, regulated and monitored than any other time in history. For example, in American Food and Drug Administration (FDA) has the primary legal responsibility for determining their safe use. To market a new food or additive (or before using an additive already approved for one use in another manner not yet approved), a manufacturer or other sponsor must first petition FDA for its approval. These petitions must provide evidence that the substance is safe for the ways in which it will be used. As a result of recent legislation, since 1999, indirect additives have been approved via a premarket notification process requiring the same data as was previously required by petition.

When evaluating the safety of a substance and whether it should be approved, the government considers:

① The composition and properties of the substance

② The amount that would typically be consumed

③ Immediate and long-term health effects

④ Various safety factors

The evaluation determines an appropriate level of use that includes a built-in safety margin - a factor that allows for uncertainty about the levels of consumption that are expected to be harmless. In other words, the levels of use that gain approval are much lower than what would be expected to have any adverse effect.

Because of inherent limitations of science, a government can never be absolutely certain of the absence of any risk from the use of any substance. Therefore, the government must determine—based on the best science available—if there is a reasonable certainty of no harm to consumers when an additive is used as proposed.

If an additive is approved, the government issues regulations that may include the types of foods in which it can be used, the maximum amounts to be used, and how it should be identifiedon food labels.

If new evidence suggests that a product already in use may be unsafe, or if consumption levels have changed enough to require another look, the government may prohibit its use or conduct further studies to determine if the use can still be considered safe.

Regulations known as Good Manufacturing Practices (GMP) limit the amount of food ingredients used in foods to the amount necessary to achieve the desired effect.

3.5 List of Food Additives on A Product Label

Food manufacturers are required to list all ingredients in the food on the label. On a product label, the ingredients are listed in order of predominance, with the ingredients used in the greatest amount first, and followed in descending order by those in smaller amounts. In America the label must list the names of any FDA-certified color additives (e. g., FD&C Blue No. 1 or the abbreviated name, Blue 1). However, some ingredients can be listed collectively as "flavors", "spices", "artificial flavoring", or in the case of color additives exempt from certification, "artificial colors", without naming each one. Declaration of an allergenic ingredient in a collective or single color, flavor, or spice could be accomplished by simply naming the allergenic ingredient in the ingredient list.

3.6 Risks of Food Additives

Despite the benefits attributed to food additives, for several years there have also been a number of concerns regarding the potential short-and long-term risks of consuming these substances. Critics of additives are concerned with both indirect and direct impacts of using additives. As for many of the benefits mentioned, there is not always adequate scientific proof of whether or not a particular additive is safe. Little or no data are available concerning the health risks or joint effects of the additive cocktail each of us consumes daily.

The indirect risks that have been described for additives are the converse of some of the benefits attributed to their use. While it is accepted that through additives a greater choice and variety of foods have been made available, there is no question that additives have also resulted in the increased availability of food products with a low density of nutrients. These so-called junk foods, which include many snack items, can in fact be used as substitutes in the diet for more nutritious foods. Recently the food industry has attempted to address this criticism by adding nutritional additives to snack items so that these foods are a source of selected vitamins and minerals. The long-term effectiveness of this is questionable. Obviously, educational programs are needed to ensure that consumers select nutritious foods. Some scientists, however, feel that there is a place in the diet for foods that provide pleasure even if no direct nutritional benefit can be ascribed to their consumption.

Of greater concern than the indirect risks are the potential direct toxicological effects of additives. Short-term acute effects from additives are unlikely. Few additives are used at levels that will cause a direct toxicological impact, although there have been incidents where this has happened. Of particular concern are the hypersensitivity reactions to some additives that can have a direct and severe impact on sensitive individuals even when the chemicals are used at legally acceptable levels. The reactions to sulfites and other additives. With proper labeling, however, sensitive individuals should be able to avoid potential allergens.

Toxicological problems resulting from the long-term consumption of additives are not well documented. Cancer and reproductive problems are of primary concern, although there is no direct evidence linking additive consumption with their occurrence in humans. There are, however, animal studies that have indicated potential problems with some additives. Although most of these additives have been banned, some continue to be used, the most notable being saccharin.

Most existing additives and all newones must undergo extensive toxicological evaluation to be approved for use. Although questions continue to be asked regarding the validity of animal studies, there is a consensus among scientists that animal testing does provide the information needed to make safety decisions.

3.7 Final Words

Food ingredients have been used for many years to preserve, flavor, blend, thicken and color foods, and have played an important role in reducing serious nutritional deficiencies among consumers. These ingredients also help ensure the availability of flavorful, nutritious, safe, convenient, colorful and affordable foods that meet consumer expectations year-round.

Food additives are strictly studied, regulated and monitored. The government require evidence that each substance is safe at its intended level of use before it may be added to foods. Furthermore, all additives are subject to ongoing safety review as scientific understanding and methods of testing continue to improve. Consumers should feel safe about the foods they eat.

References

1. 食品安全国家标准 食品添加剂使用标准（GB 2760—2014）. 中华人民共和国国家卫生和计划生育委员会发布. 2015 年 5 月 24 日起实施.

2. 食品安全国家标准 食品营养强化剂使用标准（GB 14880—2012）. 中华人民共和国卫生部发布. 2013 年 1月 1 日起实施.

3. Overview of Food Ingredients, Additives & Colors. www. fda. gov/Food/IngredientsPackagingLabeling/FoodAdditivesIngredients/ucm094211. htm＃qalabel.

4. 郝利平，聂乾忠，周爱梅. 食品添加剂（第 3 版）. 北京：中国农业大学出版社，2016.

Technical Terms

antifoaming agent 消泡剂
antioxidant ['ænti'ɔksidənt] *n*. [助剂] 抗氧化剂，抗氧化物
butylated hydroxyanisole 叔丁基羟基茴香醚
butylated hydroxytoluene 二丁基羟基甲苯
botulism ['bɒtjʊlɪz(ə)m] *n*. 肉毒中毒
color additives 着色剂
disodium 5'-inosinate 5'-肌苷酸二钠
disodium 5'-guanylate 5'-鸟苷酸二钠
dough conditioner 面团改良剂
dye [daɪ] *n*. 色素，染料
emulsifier [ɪ'mʌlsɪfaɪə] *n*. 乳化剂，黏合剂
firming agent 固化剂
flavoring agent 调味剂
food additive 食品添加剂
Food and Drug Administration of U. S. 美国食品与药物管理局
food dye 食品（用）色素
food labelling 食品标签
food preservative 食品保存剂，食品防腐剂
food regulation 食品法规
food spoilage 食物腐败，食品败坏
food supplement 食品增补剂
generally recognized as safe (GRAS)（食品添加剂的）一般公认安全，一般认为安全
humectants [hjʊ'mekt(ə)nt] *n*. 保湿剂
junket ['dʒʌŋkɪt] *n*. 凝乳食品
lake [leɪk] *n*. 色淀
leavened food 发酵食品，膨发食品

low calorie food 低热量（能量）食品
pastry ['peɪstrɪ] *n*. 焙烤（面制）食品，发面点心，面制糕点
pickled food 腌渍食品
protective food 保健食品
set milk 凝乳食品
spices [s'paɪsɪz] *n*. 香料
stabilizer ['steɪbɪlaɪzə] *n*. 稳定剂
sweetener ['swi；tənə] *n*. 甜味剂
thickener ['θɪkənə] *n*. 增稠剂
vanilla [və'nɪlə] *n*. 香草

Exercises

Answer the following questions according to the article

1. Why are food additives added to food?
2. What is a food additive?
3. How many types of common food additives are used in foods?
4. How a new additive is approved for use in foods?

Unit 4

Food Sensory Evaluation

4.1 Introduction

Of the many sectors of the consumer products industries (food and beverage, cosmetics, personal care products, fabrics and clothing, pharmaceutical, and so on), the food and beverage sectors provided much early support for and interest in sensory evaluation. During the 1940s and through the mid-1950s, sensory evaluation received additional impetus through the US Army Quartermaster Food and Container Institute, which supported research in food acceptance for the armed forces. It became apparent to the military that adequate nutrition, as measured by analysis of diets or preparation of elaborate menus, did not guarantee food acceptance by military personnel. The importance of flavor and the degree of acceptability for a particular product were acknowledged. Resources were allocated to studies of the problem of identifying what foods were more or less preferred as well as the more basic issue of the measurement of food acceptance.

The Arthur D. Little Company introduced the Flavor Profile Method, a qualitative form of descriptive analysis that minimized dependence on the technical expert. While the concept of a technical expert was and continues to be of concern, the Flavor Profile procedure replaced the individual with a group of about six experts responsible for yielding a consensus decision. This approach provoked controversy among experimental psychologists who were concerned with the concept of a group decision and the potential influence of an individual on this consensus decision. Nonetheless, at that time, the method provided a focal point for sensory evaluation, creating new interest in the discipline, which stimulated more research and development into all aspects of the sensory process.

By the mid-1950s, the University of California at Davis was offering a series of course on sensory evaluation, providing one of the new academic sources for training of sensory evaluation professionals. The early research was especially thorough in its development and evaluation of specific test methods. Discrimination test procedures were evaluated by Boggs and Hansen et al. In addition to discrimination testing, other measurement technical also were used as a means for assessing product acceptance. Rank-order procedures and hedonic scales became more common in the mid- to late 1950s. During this time period, various technical and scientific societies such as the Sensory Evaluation Division of the Institute of Food Technologists organized activities focusing on sensory evaluation and the measurement of flavor.

Certainly the international focus on food and agriculture in the mid-1960s and on into the 1970s, the energy crisis, food fabrication and the cost of raw materials, competition and internationalization of the marketplace have, directly or indirectly, created opportunities for sensory evaluation. After a long and somewhat difficult gestation, sensory evaluation has emerged as a distinct, recognized scientific specialty.

4.2 Defining Sensory Evaluation

Sensory evaluation is a scientific discipline used to evoke, measure, analyze and interpret reactions to those characteristics of foods and materials as they are perceived by the senses of sight, smell, taste, touch and hearing. Being able to identify and quantitatively model the key drivers for a products acceptance is now generally recognized as a core resource for any sensory program.

The definition makes clear that sensory evaluation encompasses all the senses. This is particularly important issue and one that is overlooked with the result that in some environments sensory evaluation is viewed solely "taste testing", as if to imply that it excludes the other senses. While an individual may be asked to respond to a particular product attribute, for example, its color, if no special care has been taken to exclude the product's aroma, then it is very likely that the obtained color response will be affected by the aroma but not in a predictable way. This will lead to a confounding of the response and potential misinterpretation of the results. A product's appearance will impact an individual's response to that product's taste, etc. Regardless of what one may like to believe or has been told responses to a product are the result of interactions of various sensory messages, independent of the source. To avoid obtaining incomplete product information, it is important to design studies that take this knowledge into account. The familiar request to "field a test but tell the subjects to ignore the color as that will be corrected later" is a sure sign of impending disaster.

The definition seeks to make clear that sensory evaluation is derived from several different disciplines, but emphasizes the behavioral basis of perception. This involvement of different disciplines may help to explain the difficulty entailed in delineating the functions of sensory evaluation within the business and academic environments. These disciplines include experimental, social, and physiological psychology, statistics, home economics, and in the case of foods, a working knowledge of food science and technology.

As the definition implies, sensory evaluation involves the measurement and evaluation of the sensory properties of foods and other materials. Sensory evaluation also involves the analysis and the interpretation of the responses by the sensory professional; that is, that individual who provides the connection between the internal world of technology and product development and the external world of the market place, within the constraints of a product marketing brief. This connection is essential such that the processing and development specialists can anticipate the impact of product changes in the marketplace. Similarly, the marketing and brand specialists must be confident that the sensory properties are consistent with the intended target and with the communication delivered to that market through advertising. They also must be confident that there are no sensory deficiencies that lead to a market failure. Linking of sensory testing with other business functions is essential just as it is essential for the sensory professional to understand the marketing strategy.

Sensory evaluation principles have their origin in physiology and psychology information derived from experiments with the senses has provided a greater appreciation for their properties, and this greater appreciation, in turn, has had a major influence on test procedures and on the measurement of human responses to stimuli. Although sources of information on sensory evaluation have improved in recently years, much information on the physiology of the senses and the behavioral aspects of the perceptual process has been available for considerably longer. As Geldard has pointed out, classically the "five special senses" are vision, audition, taste, smell, and touch. The latter designation includes the senses of temperature, pain, pressure, and so forth.

4.3 Organizing A Sensory Evaluation Program

Twelve elements that form the foundation for an effective sensory evaluation program are: stated and approved goals and objectives; defined program strategy and business plan; professional staff; test facilities, ability to use all test methods; pool of qualified subject; standardized subject-screening procedures; standardized subject performance monitoring procedures. Standardized test request and reporting procedures; on-line data processing capabilities; formal operations manual; planning and research program.

Here only facility and testing methods were introduced briefly.

4.3.1 Facilities

Basically, a certain amount of space is necessary for a sensory evaluation facility; the space requirement ranges from a minimum of 400 to 2000ft^2, However, the number of booths and number of staff to not increase in relation to the space allocation. The typical facility can be separated in six distinct areas, subject reception; six booths; preparation, holding, and storage; panel discussion; Data processing/records; Experimenter desk/office, the approximate space allocated (ft^2) are: 50, 100, 300, 350, 75, 125, respectively. The allocation of space is relative and will change depending on the types of products and the daily volume of testing. Six booths would be adequate for up to 1000 tests per year.

Lighting is the booth area is fluorescent, except in the booths themselves, where incandescent lighting is recommended. This lighting should be sufficient to provide 100-110 ft-candles (or their equivalent) of shadow-free light at the counter surface. These incandescent lights are located slightly to the front of the soffit and are titled toward the seated subject to minimize shadows on the products. Opalite diffusion glass can be placed beneath the bulbs to eliminate any spotlight effect. Off-on and dimmer switches for the booths are located in the experimenter area. The use of various types of colored light (e. g. red, yellow, and blue) for alternative booth illumination is not recommended.

4.3.2 Test Methods

In addition to an appropriate facility, sensory evaluation must have tools with which to work, including the methods used to evaluate the products (e. g. difference tests and acceptance tests). The methods are assigned to three broad categories: discriminative, descriptive, and affective, the responding test types are: difference: paired comparison, duo trio, triangle; descriptive analysis: flavor profile, QDA; and acceptance-preference: nine-point hedonic, respectively.

4.4 Guidelines for Sensory Evaluation of Meat

This guideline excerpts from "Research guidelines for cookery, sensory evaluation, and instrumental tenderness measurements of meat (1.02 version)", which was published by American Meat Science Association. It is not a "standard" to which everyone will be expected to adhere for every research study. It is, as the title suggests, "Research Guidelines." The researcher must decide the most appropriate methods to use to answer the question at hand. The methods and approaches described herein, however, are accepted and recommended as the most appropriate for most circumstances.

Recommended research methods may not always produce the highest level of consumer acceptance. The methods and approaches recommended in these Guidelines, however, are designed to control unwanted variability, to determine the most accurate answer to the questions being addressed with the most relevant methods possible, and, when feasible, to allow for valid comparative interpretation of published research.

Information is included on recommendations for collecting and preparing appropriate samples for sensory and/or tenderness evaluation for fresh beef, pork, and lamb steaks/chops, roasts, and ground patties; but it also may be applicable to certain enhanced, cured, or comminuted products. Additional topics covered include product handling, cookery methods, sensory panel methods, and a data analyses overview. Before initiating an experiment, the cooking and handling procedures, sensory method, and testing parameters should be determined.

Factors to consider in method selection include following:

- What is your hypothesis?
- What questions are you trying to answer (test objectives)?
- How will the results be used?
- How large of a difference are you trying to detect?
- How much variability is there within and between samples?

If, based on the preliminary work, the sensory differences among treatments are not expected to be detectable, the lack of significant differences can be verified using discrimination or descriptive analysis methods. If, however, the sensory differences are expected to be detectable, consumer testing methods would be more appropriate. It would be important to determine if the differences are detectable to consumers, and if detectable, how they affect consumer acceptability.

Quantitative sensory methods can be grouped into three primary categories: ①discrimination, ② descriptive analysis, and ③ consumer. Discrimination methods can use either trained panelists or untrained consumer panelists, depending on the test objectives. If the test objectives are to determine with a high degree of certainty if treatment differences are significant, trained panelists are suggested. Trained panelists are carefully selected, highly trained, and hypercritical as compared to average consumers. When using consumers for discrimination tests, consumers are not as critical and may not detect differences. The selection of a testing method should be based on the objectives of the study. Data should be interpreted based on the sample population used for the study. Descriptive methods use trained panelists. Panelists can be defined as trained (6 to 10 training sessions) to highly trained (6 months or more of training) or expert (10 or more years of experience). As the amount of training and experience increases, panelists can detect smaller differences in attributes between samples. The amount of training should be noted and data presented based on the panelists' level of training. Descriptive tests are used to quantify the level of an attribute within the samples. Depending on the testing method selected, scales and attributes may vary, but each test is used to determine if samples differ in sensory attributes. Consumer evaluation can be either qualitative or quantitative. Qualitative consumer methods provide input from consumers on their opinions on a product(s) based on more loosely structured questions that provide opportunities for consumers to give their opinions and inputs.

These data are very valuable, but they do not give quantitative results that can be statistically analyzed. Consumer quantitative testing provides an avenue to measure consumer opinions using questions on a ballot with a scale that is or can be converted to numerical values for statistical analyses. This provides a method of quantifying consumers' responses and determining differences. Test-booth conditions, coded containers, and scoring methods used in central location tests (CLTs) are certainly not typical of normal conditions of food consumption.

4.4.1 Sample Collection/Preparation

Steaks, chops, and roasts should not be sized unless preslaughter treatment or postmortem processing affects intended size. There is merit, however, in "sizing" meat cuts if variation in cooking procedures is the major focus of the project. Certainly, consideration should be given to removal of bone and connective tissue, degree of subcutaneous fat removal, thickness, weight, and shape of the cut. Ground beef patties, being thin, should have very close controls on weight and thickness. The variation in patty manufacturing parameters should be considered and standardized. All of these factors should be controlled and/or standardized to the extent necessary to collect relevant and accurate data.

(1) Selection of Samples

Researchers are strongly advised to consult with statisticians and/or perform appropriate statistical tests for determination of sample size for a research study. A number of parameters including an estimate of variability in tenderness (shear force or sensory), flavor, juiciness, or other traits of interest are needed in order to determine sample size. Many statistics textbooks provide guidelines for estimating the sample size needed for a specific experiment. It will be more fiscally and scientifically sound to have adequate sample size than to use marginal sample size and not be able to detect differences that might truly exist. These types of tests also are important for estimating the number of sensory sessions and funding to complete a project. As a rule, all samples within a study should be representative of the products or processes under study. Proper sampling within muscles is of critical importance. Steaks or chops within a muscle that are to be assigned to different treatments should be statistically randomized (if there is no known location variation) or blocked (if there is known location variation) to alleviate bias. If each carcass or cut represents one replicate of one treatment, steak location within a muscle should be standardized. For example, the first steak from the rib end of the short loin could be assigned to sensory analysis and the second steak to shear force evaluation tests. Most research that has the goal of determining production, antemortem, or postmortem treatment effects on tenderness and other palatability traits utilizes the longissimus muscle because it has the highest total value in carcasses and typically is sold as steaks or chops for dry heat cookery. Greater emphasis, however, recently has been placed on characterizing and marketing other muscles; thus, it may be appropriate to evaluate treatment effects on multiple muscles. When doing so, keep in mind that location effects are very important in some muscles. In addition, muscle selection should be made to ensure the proper answers are obtained for the experiment' s questions. Available data show that the relationships in tenderness among muscles in the same animal vary from relatively low to moderately high; thus, results in one muscle may not be representative of other muscles.

(2) Steak, Chop and Patty Variables

Following are recommended thicknesses of steaks, chops and patties:

- Beef steaks (dry heat): 2.54cm
- Beef steaks (moist heat): 1.9 to 2.54cm
- Lamb chops (dry heat): 2.54cm
- Pork chops (dry heat): 2.54cm
- Beef patties: not < 0.95cm or > 1.10cm for 91.5-g patties
- Beef patties: not < 1.10cm or > 1.27cm for 113.5-g patties

A cutting guide should be used so that steaks and chops are uniform in thickness. When product must be frozen, very uniform thicknesses can be obtained by freezing and sawing on a bandsaw. Because it is impractical to list all of the various roast cuts from all species, the reader is referred to the latest editions of the Meat Buyer' s Guide (NAMI 2015) and the Institutional meat purchase specifications (USDA, 2010) and encouraged to use common sizes found in retail and foodservice products that meet experimental objectives. The following are minimum weights and thicknesses for roasts from the three species:

- Beef: 1.5kg, 5.0cm thick
- Pork: 1.0kg, 5.0cm thick
- Lamb: 0.5kg, 5.0cm thick.

The small amount of external fat currently present on retail cuts suggests that most if not all external fat should be removed from cuts before cooking. It is recommended that products be vacuum packaged or in packaging materials with very low oxygen and moisture transmission properties.

4.4.2 Preparation and Presentation of Samples to the Panel

(1) Preparation of Sensory Samples

Selection of sample preparation method and serving size should be determined based on project objectives and the amount of variation between and within treatments. It is critical that each panelist receive a standardized amount of each sample. Standardization of samples should be not only by weight or dimensions but also by temperature.

① Trained panel evaluations

In order to account for the moderate to sometimes high degree of variability between and within treatments, meat samples often are cut into cubes, and each panelist receives two to three cubes from different locations within the piece of meat. For steaks, chops, and roasts, cubes that are 1.27cm × 1.27cm × the thickness of the cooked cut are suggested. For beef patties (depending on the size of beef patties being evaluated [91.5g or 113.5g]), cooked patties can be cut into six or eight pie-shaped samples. Even with thicker or larger-sized patties, cutting patties into cubes might result in breakage and the inability to obtain equal-sized pieces to serve the panelists. Cutting patties into pie-shaped or wedge samples is recommended.

② Consumer panel evaluations

During the normal eating experience, consumers assess the juiciness of the meat visually as well as get an initial impression of the tenderness as they cut the sample into bite-sized pieces. Therefore, when conducting consumer tests, it is best to serve samples that are large enough for the panelist to cut in order to provide a more accurate representation of the actual consumer eating experience. The serving size should be standardized. Location effects within a sub-primal should be randomized. The effect of location can then be included in the analysis of variance (ANOVA) model, usually as a random effect. Depending on the objectives and circumstances of the study, there will be times when it is advisable to serve the smaller 1.27-cm cubes. By randomizing the selection of the cubes within a steak, the variability in tenderness within a sample can be better accommodated. This results, however, in a poorer simulation of the normal eating experience when panelists receive the bite-sized cubes. Researchers should understand that they are giving up aspects of the eating experience that influence consumer perception when serving 1.27-cm cubes, and therefore, the interpretation and use of the results might be affected. The experimental design might need to be altered (i.e., increase the number of consumers per treatment) if cubes are used.

(2) Sample Presentation

Standard presentation procedures need to be followed to insure that all panelists receive samples at the most appropriate and consistent temperature for the attributes being measured. The minimum recommended serving temperature for meats is 60℃ (ASTM E1871, 2010) and under most conditions should be adhered to. For each study, procedures for holding, cutting, and serving should be determined ahead of the study. Temperature should be monitored to assure that the samples are a standard temperature when serving and that the temperature is not too high or low. Variation in temperature of samples when the panelists receive them should be known. As flavor and texture are affected by temperature and holding time, not only should the serving temperature be consistent, but the holding time of the cooked samples should be consistent as well.

Ideally, the samples will be served immediately after being cut, with panelists receiving cubes from various locations within the steak or, in the case of larger portions for consumer panelists, the same steak portion for all treatments. If samples need to be held prior to serving, preliminary tests need to be conducted to assure that holding methods do not affect color, tenderness, juiciness, or flavor of the samples. Several procedures for maintaining the temperature of the samples have been used. These include the use of the following:

- Covered pans or glass dishes that are placed in a preheated container of sand or on a warming plate or heated oven (i. e., 49℃)
- Double boilers on electric hot plates
- Wrapping the sample in aluminum foil or placing sample in Pyrex or glass baking dish with lid and storage in a heated oven
- Using preheated yogurt-maker glass dishes placed in the yogurt maker or similar apparatus

① Order of sample presentation

Every sample should be served in each serving order an equal number of times to reduce any bias related to serving position. Furthermore, every sample should be served before and after every other sample an equal number of times in order to nullify any bias related to carryover effects. William's Square designs are one way to achieve both types of balance. For some studies with a large number of treatments, this may not be possible. Order should be randomized using a random number generator, and order should be analyzed as a random effect in the model.

For trained descriptive panels, first sample bias and variation associated with evaluations conducted on different days can be an issue. A standardized warm-up sample—usually a sample that represents the typical product being evaluated in the study—should therefore be served to the panelists at the initiation of the sensory session and then discussed. This warm-up sample will assist in standardization among panelists, improve panelists' concentration, increase panelists' confidence, and increase calibration of panelists before initiating the sensory session.

The sensory leader can use the warm-up sample as a tool to address panel drift and lack of motivation, to increase panelist confidence, and to remove prior environmental factors that can influence the sensory verdict on a given day.

② Number of samples per session

The number of samples that should be presented in a given session is a function of the following:

- Product characteristics
- Experience of the panelists
- Sensory and mental fatigue
- Number of attributes to be measured per sample

Readers are encouraged to review Bohnenkamp and Berry (1987) regarding the effects of sample numbers/session and sessions/day on panelist performance in evaluating beef patties. Great care should be taken in setting the number of samples to be served per session in consumer testing. Because consumers can differ greatly in their food preferences, panelists are the greatest source of error in the statistical model. It is, therefore, best for all panelists to evaluate all products. In order to minimize taste bud fatigue and loss of interest or concentration among the panelists, however, the number of samples served per session should be limited based on the product type and the number of questions on the ballot. For unseasoned meat products, panelists can easily evaluate six to eight samples per 1-hour session. Spicy or highly seasoned products, such as marinated pork tenderloin with a spicy barbecue flavor, should be limited to four samples per session. Palate cleansers such as room temperature distilled water and unsalted crackers should also be used to minimize taste bud fatigue and flavor carryover. Additionally, fat-free ricotta cheese is an effective palate cleanser for spicy products, whereas warm water or seltzer water is effective for high-fat products. Preliminary studies should be conducted with various palate cleansers to ensure that the palate is thoroughly cleansed without contributing to taste bud fatigue or influencing attributes within the product. In situations where the number of samples in the test is greater than what each person can evaluate in one session, it is best to conduct the test over multiple days, with each panelist evaluating a subset of samples each day. If multiday sessions are not an option, then a partially balanced, incomplete-block design can be used, but the overall number of panelists should be increased to achieve the desired number of observations per sample. Consult a statistical text for tables on partially balanced, incomplete-block designs. These designs ensure that each sample or treatment appears with each other sample or treatment an equal number of times in any given session.

(3) Sensory Panel Participants' Informed Consent

The Belmont Report entitled Ethical principles and guidelines for the protection of human subjects of research was created by the Department of Health, Education, and Welfare in 1979 to safeguard human subjects used for research, including sensory panels (NIH, 1979). This report protects human subjects used in research by fulfilling three fundamental ethical principles:

- Respect for persons—protecting the autonomy of all people and treating them with courtesy and respect and allowing for informed consent
- Beneficence—maximizing benefits for the research project while minimizing risk to the research subjects
- Justice—ensuring that reasonable, non-exploitative, and well-considered procedures are administrated fairly

In the United States, federally funded research and research conducted at state and federally supported institutions involving human subjects such as sensory panels must obey ethical rules that include obtaining participants' informed consent and supervision by an Institutional Review Board (IRB). The form prepared for obtaining participants' consent should precisely convey all the information about the sensory panel to the participants. The consent form should include the purpose of study and study design, who can participate, who will be conducting the research, what participants will be asked to do, possible risk and discomforts, possible benefits/compensation.

Technical Terms

cosmetics [kɑz'mɛtɪks] *n*. 化妆品（cosmetic 的复数）；装饰品

pharmaceutical [ˌfɑrmə'sutɪkl] *adj*. 制药（学）的 *n*. 药物

impetus ['ɪmpɪtəs] *n*. 动力；促进；冲力

evoke [ɪ'vok] *vt*. 引起，唤起；博得

stimuli ['stɪmjʊlaɪ] *n*. 刺激；刺激物；促进因素（stimulus 的复数）

applicable ['æplɪkəbl] *adj*. 可适用的；可应用的；合适的

comminuted ['kɑməˌnjʊt] *adj*. 粉碎的 *v*. 粉碎（comminute 的过去分词）；使成粉末

hypercritical ['haɪpɑ'krɪtɪkl] *adj*. 吹毛求疵的，苛评的

variation [ˌvɛrɪ'eʃən] *n*. 变化，变异，变种

impractical [ɪm'pæktɪkl] *adj*. 不切实际的，不现实的，不能实行的

marinate ['mærɪnet] *vt*. 把……浸泡在卤汁中 *vi*. 浸泡在卤汁中

palate ['pælət] *n*. 味觉，上颚，趣味

Exercises

Ⅰ. Answer the following questions according to the article

1. Please define the sensory evaluation.

2. Please list the categories of tests and examples of methods used in sensory evaluation.

3. Before initiating an experiment, the cooking and handling procedures, sensory method, and testing parameters should be determined. What factors to consider in method selection?

4. What are the recommended thickness of steaks, chops and patties?

5. What are the rules for the order of sample presentation?

Ⅱ. Choose a term from what we have learnt to fill in each of the following blanks. Change the word form where necessary.

1. ____________________ for a products acceptance is now generally recognized as a core resource for any sensory program.

2. The definition seeks to make clear that sensory evaluation is derived from several different disciplines, but emphasizes ____________________.

3. Lighting is the booth area is __________, except in the booths themselves, where ____________ is recommended.

4. This manual is not a "standard" to which everyone will be expected to adhere for every research study. It is, as the title suggests, ________________.

5. For steaks, chops, and roasts, cubes that are ____________________ of the cooked cut are suggested.

6. A standardized warm-up sample should ____________________ therefore be served to the panelists at the initiation of the sensory session and then discussed.

7. For unseasoned meat products, panelists can easily evaluate ____________ samples per 1-hour session.

8. Palate cleansers such as ____________________ should also be used to minimize taste

bud fatigue and flavor carryover.

9. The form prepared for obtaining participants' consent should include the purpose of study and study design, ________________ possible risk and discomforts, possible benefits/compensation.

Unit 5

Food Toxicology

5.1 Toxic Agents

One could define a poison as any agent capable of producing a deleterious response in a biological system, seriously injuring function or producing death. This is not a useful working definition, however, for the simple reason that virtually every known chemical has the potential to produce injury or death if it is present in a sufficient amount (dose). At sufficiently high doses, any chemical may be described as toxic. The importance of dose is clearly seen with molecular oxygen or dietary metals. Oxygen at a concentration of 21% in the atmosphere is essential for life, but 100% oxygen at atmospheric pressure causes massive lung injury in rodents and often results in death. Some metals such as iron, copper, and zinc are essential nutrients. When they are present in insufficient amounts in the human diet, specific disease patterns would develop, but in high doses they can cause fatal intoxications.

Therefore, all toxic effects are products of the amount of chemical to which the organism is exposed and the inherent toxicity of the chemical; they also depend on the sensitivity of the biological system. The toxicity of a chemical could be defined as the ability to cause a deleterious effect to a biological system. Among chemicals there is a wide spectrum of doses needed to produce toxic effects.

5.2 Dose and Concentration

The most important factor influencing the potential toxicity of a chemical is the dose or concentration. Remember, anything can be toxic at a given dose. Conversely, even the most toxic substances may not be harmful or even be necessary for good health at extremely low concentrations. Water, the fluid of life, is considered completely nontoxic, acutely and chronically. Nevertheless, at a high enough volume or dose, its toxicity becomes evident. A few cases have been reported in the medical literature of both acute and chronic intoxications, some even fatal, from excessive water intake. Quantities considered excessive are measured in gallons per day. Death from drinking excessive amounts of water occurs as a consequence of literally drowning the cells and tissues of the body. Another example is sodium fluoride. Sodium fluoride is acutely highly toxic but useful in trace amounts. It has an oral LD_{50} of about 35 mg/kg. Chronically, in very small amounts of 1-2 mg daily, it is needed for good dental health. In quantities of 3mg or 4mg per day or greater, sodium fluoride can cause mottling of tooth enamel in young people. Larger daily quantities can produce chronic fluorosis, a condition characterized by increased bone density and the formation of bone spurs. Vitamin D in pure form is highly toxic acutely, with an oral LD_{50} of about 10mg/kg, or 400000IU/kg, the same as a parathion. Despite its high acute toxicity, an average of 10pg (400IU) of vitamin D every day is required for our good health. Vitamin D deficiency results in the disease known as rickets. Severe vitamin D deficiency can eventually cause death.

5.3 Detinition of Toxicology

Toxicology is the study of the adverse effects of chemicals on living organisms. It is a multidisciplinary subject which comprises many different areas. Apart from the specialization within toxicology, a toxicologist is trained to perform one or both of the two basic functions of toxicology, which are to ①examine the nature of the adverse effects produced by a chemical and ②assess the probability of these hazards/toxicities occurring under specific conditions of exposure. Ultimately, the goal and basic purpose of toxicology is to provide a basis for appropriate controlling measures so that these adverse effects can be prevented.

5.4 Scope of Toxicology

5.4.1 Descriptive Toxicology

A descriptive toxicologist is concerned directly with toxicity testing, which provides information of safety evaluation and regulatory requirements. The concern may be limited to effects on humans, in the case of drugs and food additives, and potential effects on fish, birds, and plants, as well as other factors that might disturb the balance of the ecosystem.

5.4.2 Mechanistic Toxicology

A mechanistic toxicologist is concerned with identifying and understanding the mechanisms through which chemicals exert toxic effects on living organisms. In risk assessment, mechanistic data can be very useful in demonstrating an adverse outcome observed in laboratory animals, which is or is not directly relevant to humans. Mechanistic data are also useful in the design and production of safer alternative chemicals and in rational therapy for chemical poisoning and treatment of disease. An understanding of the mechanisms of toxic action contributes to the knowledge of basic physiology, pharmacology, cell biology, and biochemistry.

5.4.3 Regulatory Toxicology

A regulatory toxicologist has the responsibility to decide, on the basis of data provided by descriptive and mechanistic toxicologists, whether a drug or another chemical poses a low enough risk to be marketed for a stated purpose. Regulatory toxicologists also assist in the establishment of standards for the amount of chemicals permitted in ambient air, industrial atmospheres, and drinking water. They often integrate scientific information from basic descriptive and mechanistic toxicology studies with the principles and approaches used for risk assessments.

5.5 Methods and Types of Toxicology Studies

It is, of course, best to assess the potential human hazard of a particular chemical depending on human data that have been generated for the same exposure conditions. Unfortunately, such data are rarely available. The most typical human data available are generated from human populations in some occupational or clinical settings in which the exposure was believed, at least initially, to be safe, except for those infrequent, unintended poisonings and environmental releases. Thus, data may come from up to four or five different categories of toxicology test of the safety evaluation of a particular chemical. These categories are: basic animal toxicology tests, the less traditional alternative tests (mainly, *in vitro* methods), epidemiology studies, clinical exposure studies, and accidental acute poisonings.

Each type or category of toxicology study has its own advantages and disadvantages and no single test system is likely to be ideal when it's used to assess the potential human hazard or safety of a particular chemical. Therefore, it is necessary to weigh the strengths and weaknesses of each test system in order to reach a conclusion. These have been summarized in Table 5.1, which lists some of the advantages and disadvantages of toxicology studies of each category.

Table 5.1 Some advantages and disadvantages of toxicology studies by category

Type of study	Advantages	Disadvantages
In vivo methods	①Easily manipulated and controlled ② Evaluated effects on intact animals and assess organ system interactions ③Widest range of potential effects to study ④The chance to identify and elucidate mechanisms of adverse effects	①Response of the test species is of uncertain human relevance ②Exposure levels may not be relevant to the human exposure level ③ Structural and biochemical differences between test animals and humans, which make extrapolations from one to the other difficult
In vitro methods	①Less expensive and quick to perform ②Easier to control host factors ③Conservation of animal resources and easier to be accepted ethically ④Possible to use human tissues	①Cannot fully approximate the complexities that take place in the whole organism ② Inability to detect delayed and/or chronic toxic effects
Epidemiological studies	① The direct observation of effects in humans ② Real-life exposure conditions relevant to chemical-induced health effects ③ The full range of human susceptibility may be measurable ④The chance to study the interactive effects of other chemicals	①Often costly and time-consuming ②Many confounding risk factors are present ③Exposures may have been poorly documented ④ Determining crude endpoints of exposure such as mortality and morbidity ⑤ Not necessarily designed to be protective of health
Clinical (human) exposure studies	①The conditions of these studies are better defined and controlled ②Most relevant species (humans) to study ③ Exposure conditions may be altered during the exposure interval in response to the presence or lack of an effect making NOAEL easier to obtain	①May be costly to perform ② The most sensitive group (e.g., young, elderly, infirm) may often be inappropriate for study ③ Primarily limited to examining safe exposure levels and minimally serious effects ④ Usually limited to shorter exposure intervals

续表

Type of study	Advantages	Disadvantages
Accidental acute poisonings	①Exposure conditions are realistic ②Require very few individuals to perform ③Inexpensive than other human studies	①Accurate exposure information may be lacking ② The knowledge gained from these studies may be of limited relevance to other human exposure situations

5.6 Dose, Response, and Dose-response Relationship

5.6.1 Dose

Dose usually implies the exposure dose, the total amount of chemical administered to an organism or incorporated into a test system. It usually appears in the form of quantity per unit body weight given on a single occasion (g/kg) or repeated daily [g/(kg · d)]. However, dose may not be directly proportional to the toxic effects since toxicity depends on the amount of chemical absorb. The concept of dose can be further defined in terms of internal dose, delivered dose and biologically effective dose. Internal (absorbed) dose is the actual quantity of a toxicant that is absorbed into the organism and distributed systemically throughout the body. Delivered (target organ) dose refers to the amount of toxicant that gets to the target organ, causing a measurable effect. The amount of an internal dose necessary to elicit a response or health effect is referred to as the biologically effective dose.

In order for the word dose to be meaningful, the route, duration and frequency of exposure also should be indicated. The major routes in which toxic agents gain access to the body are the gastrointestinal tract, lungs, skin, and other parenteral routes. The exposure of animals to chemicals is usually divided into four categories: acute exposure, subacute exposure, subchronic exposure, and chronic exposure. An acute exposure is defined as the exposure to a chemical for less than 24h. A subacute exposure refers to repeated exposure to a chemical for 1 month or less, whereas subchronic for 1 to 3 months, and chronic for more than 3 months.

5.6.2 Two Different Types of Response—graded and Quantal

From a practical point of view, there are two different types of response: graded and quantal. The graded (or variable) response involves a continual change in effects with increasing dose, for example, effects on body weight, food consumption, enzyme inhibition and changes in physiological function such as heart rate. Graded responses may be determined in an individual or in simple biochemical systems. For example, addition of increasing concentrations of 2,3,7,8-tetrachlorodibenzodioxin to cultured mammalian cells results in an increase in the concentration of a specific cytochrome P450 enzyme in the cells. Low doses of the toxin cause slight irritation in the cells, however, as the amount increases, the irritation turns to inflammation and the severity of inflammation increases. On the other hand, mortality and tumor formation are examples of quantal (all-or-none) responses in which the defined effect is either present or absent.

5.6.3 The Concept of Dose-response Relationship

The characteristics of exposure and the spectrum of effects come together in a correlative relationship customarily referred to as the dose-response relationship. The concept in its simplest form states that as one modify the 'dose' of potentially toxic material to which a living organism is exposed, the 'response' will also be modified. Whatever response is selected for measurement, the dose-response relationship assumes a form that occurs so consistently as to be considered as the most classic and fundamental concept in toxicology. Indeed, an understanding of this relationship is essential to the study of toxic chemicals.

5.7 Summary

Toxicology is multidisciplinary subject, which is the study of the adverse effects of chemicals on living organisms. The two basic functions of toxicology are to ①examine the nature of the adverse effects produced by a chemical and ②assess the probability of these hazards/toxicities occurring under specific conditions of exposure. At sufficiently high doses, any substance may be described as toxic.

Technical Terms

deleterious [ˌdelə'tɪriəs] *adj*. 有害的；有毒的，harmful
virtually ['vɜ：rtʃuəli] *adv*. 无形；无形中；实际上，实质上，事实上，几乎
dose [doʊs] *n*. 剂量
iron ['aɪərn] *n*. 铁
copper ['kɑ：pə (r)] *n*. 铜
zinc [zɪŋk] *n*. 锌
intoxication [ɪnˌtɑksɪ'keʃən] *n*. 中毒
organism ['ɔ：rgənɪzəm] *n*. 有机体；生物体；微生物；有机体系，有机组织
toxicity [tɑ：k'sɪsəti] *n*. 毒性 tox-：毒，有毒
toxic ['tɑ：ksɪk] *adj*. 有毒的
toxicology [ˌtɑ：ksɪ'kɒlədʒi] *n*. 毒理学
toxicant ['tɒksɪkənt] *n*. 毒物
toxicologist [ˌtɒksɪ'kɒlədʒɪst] *n*. 毒理学工作者，毒理学家
toxicological [ˌtɒksɪkə'lɒdʒɪkl] *adj*. 毒理学的
toxicosis [ˌtɒksɪ'koʊsɪs] *n*. 中毒
toxin ['tɑ：ksɪn] *n*. 毒素
spectrum ['spɛktrəm] *n*. 光谱；波谱；范围；系列
synthetic [sɪn'θɛtɪk] *adj*. 合成的；人造的；模拟的，虚构的；*n*. 合成物；合成纤维；合成剂
anthropogenic [ˌænθrəpə'dʒnɪk] *adj*. 人类起源论的；人为的，人类活动产生的
acute [ə'kjut] *adj*. 尖的，锐的；敏锐的，敏感的；严重的，剧烈的；[医] 急性的
acutely [ə'kju：tli] *adv*. 尖锐地；剧烈地
chronic ['krɑ：nɪk] *adj*. 慢性的；长期的；习惯性的
chronically ['krɒnɪklɪ] *adv*. 慢性地，习惯性地，长期地
sodium fluoride ['sodiəm 'flʊrˌaɪd] *n*. 氟化钠
mottle ['mɑ：tl] *v*. 杂色，斑点
enamel [ɪ'næməl] *n*. 搪瓷；珐琅；指甲油；*vt*. 给……上珐琅；在……涂瓷漆

fluorosis [ˌfluːəˈroʊsɪs] *n*. 氟中毒

bone spurs *n*. 骨刺

parathion [ˌpærəˈθaɪɒn] *n*. 对硫磷

rickets [ˈrɪkɪts] *n*. 软骨病，佝偻病，驼背

multidisciplinary [ˌmʌltidɪsəˈplɪnəri] *adj*. 多学科；包括各种学科的，有关各种学问的；多部门；多科目

pharmacology [ˌfɑːrməˈkɑːlədʒi] *n*. 药理学，药物学

epidemiology [ˌepɪdiːmiˈɑːlədʒi] *n*. 流行病学

ethically [ˈeθɪklɪ] *adv*. 伦理（学）上

endpoint [ˈendˌpɔɪnt] *n*. 端点，终点

mortality [mɔːrˈtæləti] *n*. 死亡数，死亡率；必死性，必死的命运

morbidity [mɔːˈbɪdətɪ] *n*. 发病率，发病，病态，不健全

gastrointestinal [ˌgæstroʊɪnˈtestɪnl] *adj*. 胃肠道的

parenteral [pəˈrentərəl] *adj*. 肠胃外的；不经肠的；非肠道的；注射用药物的

subacute [ˌsʌbəˈkjuːt] *adj*. 亚急性的

subchronic [sʌbkˈrɒnɪk] *adj*. 亚慢性的

quantal [ˈkwɑntl] *adj*.（实验中）仅有两种可能的；量子的，量子论的

2,3,7,8-tetrachlorodibenzodioxin [tetrəklɒroʊdɪˈbenzoʊdɪ] *n*. 化合物，2,3,7,8-四氯二苯-对二噁英

mammalian [mæˈmeljən] *adj*. 哺乳动物的

cytochrome [ˈsaɪtəˌkroʊm] *n*. 细胞色素

irritation [ˌɪrɪˈteʃən] *n*. 刺激；激怒，恼怒，生气；兴奋；令人恼火的事

inflammation [ˌɪnfləˈmeʃən] *n*. 炎症

Exercises

Ⅰ. Answer the following questions according to the article

1. What is the definition of toxicology?
2. What are the three essential elements of toxicology?
3. What are the three main specialized areas of toxicology?
4. How many types do toxicology studies have? What are they?
5. What is the meaning of LD_{50}?

Ⅱ. Choose a term from what we have learnt to fill in each of the following blanks. Change the word form where necessary

1. Toxicology can be defined as the study of ______________.
2. A toxic substance produced by biological systems is specifically referred to as a __________.
3. ________ is the most important in determining the extent of toxicity of a chemical.
4. The response of an individual to varying doses of a chemical is often referred to as ________.
5. ________ is important for setting a safe dose (e. g., ADI, RfD).

Unit 6

Food Packaging

6.1 Introduction

Nowadays food packaging plays an important role in food industry. The packs have multifunctions; they contain, preserve and protect the product. The outer covering should also to inform the consumer about the product and the design should promote the product. The packaging also has a secondary function, that of loss, damage and waste reduction for distributor and customer and the facilitation of storage, handling and other commercial operations.

Packaging technologies bring together a vast range of techniques and materials with two basic objectives: to protect the product and to display items for sale. Therefore, packaging has thus progressed from the functional to the expressive as result of some factors: the aim to motivate customers to buy the product and to convey a suitable product image for selling.

According to the FAO report, 50% of agricultural produces are lost because of the absence of packaging. The causes of this loss are bad weather, physical, chemical and microbiological deteriorations. Industrialization and the consumption of natural resources has accelerated progress so the manufacturer of packaging or related machines have to adapt and to anticipate trends and realize that only automation can provide the necessary flexibility to satisfy industrial needs. Progress in the packaging of foodstuffs will be crucial over the next few years mainly because of new consumer patterns and demands creation and of world population growth which is estimated to 15 billion by 2025.

6.2 Functions of Food Packaging

The functions of packaging are numerous and include such purposes as protecting raw or processed foods from spoilage and contamination by an array of external hazards. Packaging serves as a barrier in controlling potential damaging levels of light, oxygen, and water. It facilitates ease of use, offers adequate storage, conveys information, and provides evidence of possible product tampering. It achieves these goals by the following manners:

- Preserving against spoilage of color, flavor, odor, texture, and other food qualities.
- Preventing contamination by biological, chemical, or physical hazards.
- Controlling absorption of O_2 and losses and water vapor.
- Facilitating ease of using product contents, such as packaging that incorporates the

components of a meal together in meal "kits" (eg, tacos).

• Offering adequate storage before use, such as stackable, resealable, and pourable.

• Preventing/indicating tampering with contents by tamper-evident labels.

• Communicating information regarding ingredients, nutrition facts, manufacturer name and address, weight, bar code information, and so forth via package labeling.

• Marketing, standards of packaging, including worldwide acceptability of certain colors and picture symbols vary and should be known by the processor.

Packages themselves may promote sales. They may be rigid, flexible, metalized, and so forth, and they may carry such information as merchandising messages, health messages, recipes, and coupons.

6.3 Food Packaging Techniques

In the last decade, the most important additional function of the packaging method was to prolong the shelf-life of the food product. There are many new methods, used worldwide, in food packaging based on reduced -oxygen atmosphere surrounding the product.

(1) Vacuum Packaging

Vacuum packaging involves the placing, either manually or automatically, of a perishable food inside a plastic film package and then, by physical or mechanical means, removing air especially oxygen (O_2) from inside the package so that the packaging material remains in close contact with the product surfaces after sealing.

Packaging in this manner, depending on the product being packaged the barrier properties of the packaging material, the level of air removal, and storage temperature, can substantially retard chemical and/or microbial deterioration of the food product. In many instances, this dramatically extends eating quality life.

(2) Active Packaging

Active packaging has been defined as "packaging in which subsidiary constituents have been deliberately included in or on either the packaging material or the package headspace to enhance the performance of the package system". Active packaging includes additives or freshness enhancers that are capable of scavenging oxygen; adsorbing carbon dioxide, moisture, ethylene and/or flavor/odor taints; releasing ethanol, sorbates, antioxidants and/or other preservatives; and/or maintaining temperature control.

(3) Intelligent Packaging

Intelligent or smart packaging is not synonymous with active packaging. It refers to packaging that senses and informs. Intelligent packaging devices are capable of sensing and providing information about the function and properties of packaged food and can provide assurances of pack integrity, tamper evidence, product safety and quality, and are being utilized in applications such as product authenticity, anti-theft and product traceability. Intelligent packaging devices include time-temperature indicators (TTI), gas sensing dyes, microbial growth indicators, physical shock indicators, and numerous examples of tamper proof, anti-counterfeiting and anti-theft technologies.

(4) Edible Coatings

The technique uses edible coatings or films, which can act to food product as a protective superficial layer i. e. for many years waxing of fruits. Currently, edible films and coatings which protect a food against microbial spoilage as well as loss of quality are developed on the basis of proteins, starches, waxes, lipids, antimicrobial and antioxidant compounds.

(5) Modified Atmosphere Packaging (MAP)

MAP means that an atmosphere with a gas composition different from that of atmospheric air is created in the package. The properties of the main used gases are the following: CO_2—antimicrobial effect. O_2—for most of the processed foods, the packaging should reduce the oxygen concentration in the headspace of a package under 1%～2%, even to 0. 2%. While, for some fresh agricultural products, a certain amount of oxygen is needed in order to extend the shelf-life. N_2—inert gas.

In MAP of non-respiring foods (no need for O_2) a high CO_2 content (>20%) is used in most cases with a low O_2 content (<0. 5%) and a storage temperature <5℃ is recommended. In MAP of respiring foods, i. e. fresh fruits and vegetables, once the atmosphere has been changed to the desired level, the respiration rate of the produce should equal the diffusion of gases across the packaging material in order to maintain an equilibrium atmosphere in the package.

(6) Aseptic Packaging

Aseptic packaging normally means that foods after heat processing are transferred to "sterile" and hermetically sealed containers under aseptic conditions to avoid re-infection form taking place. The principle is well known for liquid products, e. g. (UHT) milk, fruit juices etc. Independent sterilization of both the foods and packaging material, with assembly under sterile environmental conditions, is the rule for aseptic packaging that now shows more mainstream technology.

In an aseptic system of packaging, the packaging material consists of layers of polyethylene, paperboard, and foil. It is sterilized by heat (superheated steam or dry hot air) or a combination of heat and hydrogen peroxide and then roll-fed through the packer to create the typical brick/block shape.

6.4 Selection of Food Packaging Materials

For choosing the appropriate packaging for their products, packers must consider many variables. For example, canners must make packaging choices based on cost, product compatibility, shelf-life, flexibility of size, handling systems, production line filling and closing speeds, processing reaction, impermeability, dent and tamper resistance, and consumer convenience and preference.

Processors who use films for their products must select film material based on its "barrier" properties that prevent oxygen, water vapor, or light from negatively affecting the food. As an example, the use of packaging material that prevents light-induced reactions will control degradation of the chlorophyll pigment, bleaching or discoloration of vegetable and red meats, destruction of riboflavin in milk, and oxidation of vitamin C. Some films are heat stable for cooking applications, and some show cold temperature resistance in refrigerated or frozen storage. The most common food packaging materials include metals, glass, paper, and plastic.

(1) Metal

Metals such as steel and aluminum are used in cans and trays. A metal can forms a hermetic seal, which is a complete seal against gases and vapor entry or escape and it offers protection to the contents. The trays may be reusable or disposable, recyclable trays and either steam table or No. 10 can size. Metal is also used as bottle closures and wraps.

(2) Glass

Glass is derived from metal oxides such as silicon dioxide (sand). It is used in forming bottles or jars (which subsequently receive hermetic seals), and thus protects against water vapor or oxygen loss. The thickness of glass must be sufficient to prevent breakage from internal pressure, external impact, and thermal stress. Glass that is too thick increases weight, and thus freight costs, and is subject to an increased likelihood of thermal stress or external impact breakage.

(3) Paper

Paper is derived from the pulp of wood and may contain additives such as aluminum particle laminates, plastic coating, resins, or waxes. These additives provide burst strength (strength against bursting), wet strength (leak protection), and grease and tear resistance, as well as barrier properties that assure freshness, protect the packaged food against vapor loss and environmental contaminants, and increase shelf-life.

Varying thicknesses of paper may be used to achieve thicker and more rigid packaging.

- Paper is thin (one layer) and flexible, typically used in bags and wrappers. Kraft (or "strong" in German) paper is the strongest paper. It may be bleached and used as butcher wrap or may remain unbleached and used in grocery bags.
- Paperboard is thicker (although still one layer) and more rigid. Ovenable paperboard is made for use in either conventional or microwave ovens by coating paperboard with PET polyester (see Plastic).
- Multilayers of paper form fiberboard, which is recognized as "cardboard."

(4) Plastic

Plastic has shrink, nonshrink, flexible, semirigid, and rigid applications, and varies in its degree of thickness. Important properties of the many types of plastics that make them good choices for packaging material include the following:

- Flexible and stretchable;
- Lightweight;
- Low-temperature formability;
- Resistant to breakage, with high burst strength;
- Strong heat sealability;
- Versatile in its barrier properties to O_2, moisture, and light.

The main plastic food packaging materials are polyethylene (PE), polypropylene (PP), polyethylene terephthalate (PET), polystyrene (PS), polyvinyl chloride (PVC), polyamide

(PA), polyvinylidene chloride (PVDC) and ethylene vinyl alcohol (EVOH).

PE is the most common and the least expensive plastic, comprising 63% of total plastic packaging. It is a water-vapor (moisture) barrier and prevents dehydration and freezer burn. PP has a higher melting point and greater tensile strength than PE. It is often used as the inside layer of food packages that are subject to higher temperatures of sterilization (eg, retort pouches or tubs). PET is used in "an increasing number of food and beverages", including used as a tube which dispenses food. Some advantages of PET are that it withstands high temperature foods and is lighter in weight than the glass that it replaces. PS is a versatile, inexpensive packaging material and represents 8% of total plastic packaging. When foamed, its generic name is expandable polystyrene (EPS). This styrofoam has applications in disposable packaging and drinking cups. It offers thermal insulation and protective packaging. PVC comprises 6% of total plastic packaging. It blocks out air and moisture, preventing freezer burn, and offers low permeability to gases, liquid, flavors, and odors. PVC prevents the transfer of odor and keeps food fresh by controlling dehydration and is capable of withstanding high temperatures without melting. PA belongs to a family of polymers obtained by condensation of monomers: di-amines and bi-carboxylic acids, or amino acids that have both functional ends in the same molecule. PA properties can vary in a broad range, according to its MW and crystallinity. In general, these polymers have good gas barrier, puncture resistance, and heat resistance properties. PVDC, a copolymer of vinylidene chloride (85%~90%) and vinyl chloride, is commercialized under the trade name "Saran". The most notable advantages of PVDC are related to its excellent oxygen and moisture barriers. It is mostly used in multilayer films and containers coating. EVOH which also has very good performance as oxygen barrier. In addition, it is much more common in multilayer structures. Depending on molar proportions of ethylene and vinyl alcohol in the copolymer, barrier properties can change hugely.

6.5 Future Trends in Food Packaging

A continuing trend in food packaging technology is the study and development of new materials that possess very high barrier properties. High-barrier materials can reduce the total amount of packaging materials required, since they are made of thin or lightweight materials with high-barrier properties. Convenience is also a "hot" trend in food packaging development. Convenience at the manufacturing, distribution, transportation, sales, marketing, consumption and waste disposal levels is very important and competitive. A third important trend is safety, which is related to public health and to security against bioterrorism. It is particularly important because of the increase in the consumption of ready-to-eat products, minimally processed foods and pre-cut fruits and vegetables. Food-borne illnesses and malicious alteration of foods must be eliminated from the food chain.

Food science and packaging technologies are linked to engineering developments and consumer studies. Consumers tend continuously to want new materials with new functions. New food packaging systems are therefore related to the development of food-processing technology, lifestyle changes, and political decision-making processes, as well as scientific confirmation.

Technical Terms

Food and Agriculture Organization (FAO) 联合国粮食与农业组织
vacuum packaging 真空包装
shelf-life 货架寿命
active packaging 活性包装
intelligent packaging 智能包装
modified atmosphere packaging 改善气氛包装（气调包装）
aseptic packaging 无菌包装
edible coatings 可食性涂层
tamper-evident labels 防篡改标签（显窃启标签）
oxygen scavengers 氧气脱除剂（脱氧剂）
CO_2 emitters 二氧化碳发生剂（二氧化碳释放剂）
barrier properties 阻隔性能
burst strength 破裂强度（耐破度）

Exercises

Ⅰ. Answer the following questions according to the article

1. What are the functions of food packaging?
2. How to choose proper packaging materials for food?
3. Please list the types of food packaging and analyze the difference between them.
4. Give an example of food packaging and discuss the requirements for packaging.
5. What are the effects of CO_2, O_2 and N_2 in MAP?
6. Please describe the characteristics of plastic packaging materials.
7. Discuss the future trends in food packaging.

Ⅱ. Choose a term from what we have learnt to fill in each of the following blanks. Change the word form where necessary

1. Food packaging can the ________, ________ and ________ product.
2. The causes of agricultural products' loss are ________, ________, ________ and ________.
3. Active packaging can change ________ in the package by placing sachets with ________ or ________ in the package or using another special means.
4. The barrier properties of food packaging films often refer to ________, ________ and ________.
5. When choosing the appropriate packaging material for their product, properties such as ________, ________, ________ and ________ and so forth should be considered.
6. ________, ________ and ________ show excellent oxygen barrier performance in plastic materials.

Unit 7

Sanitation in Food Plant

Food sanitation is defined as the "hygienic practices designed to maintain a clean and wholesome environment for food production, preparation and storage".

Sanitary design of food production, processing equipment, etc., is the most important factor in ensuring that food is safe and wholesome.

7.1 Plant Construction and Design

Plant buildings and structures shall be suitable in size, construction, and design to facilitate maintenance and sanitary operations for food-manufacturing purposes. The plant and facilities shall:

① Provide sufficient space for such placement of equipment and storage of materials as is necessary for the maintenance of sanitary operations and the production of safe food.

② Permit the taking of proper precautions to reduce the potential for contamination of food, food-contact surfaces, or food-packaging materials with microorganisms, chemicals, filth, or other extraneous material. The potential for contamination may be reduced by adequate food safety controls and operating practices or effective design, including the separation of operations in which contamination is likely to occur, by one or more of the following means: location, time, partition, airflow, enclosed systems, or other effective means.

③ Permit the taking of proper precautions to protect food in outdoor bulk fermentation vessels by any effective means.

④ Be constructed in such a manner that floors, walls, and ceilings may be adequately cleaned and kept clean and kept in good repair.

⑤ Provide adequate ventilation or control equipment to minimize odors and vapors (including steam and noxious fumes) in areas where they may contaminate food.

7.2 Sanitary Operations

① General maintenance. Buildings, fixtures, and other physical facilities of the plant shall be maintained in a sanitary condition and shall be kept in repair sufficient to prevent food from becoming adulterated within the meaning of the act.

② Substances used in cleaning and sanitizing; storage of toxic materials: Cleaning compounds and sanitizing agents used in cleaning and sanitizing procedures shall be free from undesirable microorganisms and shall be safe and adequate under the conditions of use. Toxic cleaning compounds, sanitizing agents, and pesticide chemicals shall be identified, held, and stored in a manner that protects against contamination of food, food-contact surfaces, or food-packaging materials.

③ Pest control. No pests shall be allowed in any area of a food plant. Guard or guide dogs may be allowed in some areas of a plant if the presence of the dogs is unlikely to result in contamination of food, food-contact surfaces, or food-packaging materials.

④ Sanitation of food-contact surfaces. All food-contact surfaces, including utensils and food-contact surfaces of equipment, shall be cleaned as frequently as necessary to protect against contamination of food. Non-food-contact surfaces of equipment used in the operation of food plants should be cleaned as frequently as necessary to protect against contamination of food. Single-service articles (such as utensils intended for one-time use, paper cups, and paper towels) should be stored in appropriate containers and shall be handled, dispensed, used, and disposed of in a manner that protects against contamination of food or food-contact surfaces.

7.3 Sanitary Facilities and Controls

Each plant shall be equipped with adequate sanitary facilities and accommodations including, but not limited to:

① Water supply. The water supply shall be sufficient for the operations intended and shall be derived from an adequate source.

② Plumbing. Plumbing shall be of adequate size and design and adequately installed and maintained to carry sufficient quantities of water to required locations throughout the plant, avoid constituting a source of contamination to food, water supplies, equipment, or utensils or creating an unsanitary condition.

③ Sewage disposal. Sewage disposal shall be made into an adequate sewerage system or disposed of through other adequate means.

④ Toilet facilities. Each plant shall provide its employees with adequate, readily accessible toilet facilities.

⑤ Hand-washing facilities. Hand-washing facilities shall be adequate and convenient and be furnished with running water at a suitable temperature. Compliance with this requirement may be accomplished by providing:

Effective hand-cleaning and sanitizing preparations;

Sanitary towel service or suitable drying devices.

Technical Terms

sanitation [ˌsænɪ'teɪʃn] *n*. 卫生系统或设备

sanitary ['sænətri] *adj*. 清洁的；卫生的；*n*. 公共厕所

storage ['stɔːrɪdʒ] *n*. 储存；储藏；储藏处，仓库；储存器

facilitate [fə'sɪlɪteɪt] *vt*. 促进，助长；使容易；帮助

facilities [fə'sɪlɪtɪz] *n*. 工具；天资；（机器等的）特别装置；设备（facility 的名词复

数）；能力

precautions [pri'kɔːʃənz] *n*. 预防措施（ precaution 的名词复数）；防备；避孕措施

microorganism [ˌmaɪkrəʊ'ɔːgənɪzəm] *n*. 微生物

fermentation [ˌfɜːmen'teɪʃn] *n*. 发酵；激动，纷扰

ventilation [ˌventɪ'leɪʃn] *n*. 空气流通；通风设备；通风方法；公开讨论

maintenance ['meɪntənəns] *n*. 维持，保持；保养，保管；维护

articles ['ɑːtiklz] *n*. 用品；条款（article 的名词复数）；物品；（报刊上的）论文

single-service articles 一次性用品

requirement [rɪ'kwaɪəmənt] *n*. 要求；必要条件；必需品，需要量；资格

construction [kən'strʌkʃn] *n*. 建筑物；建造；解释；建造物

manufacturing [ˌmænju'fæktʃərɪŋ] *n*. 制造业；*adj*. 制造业的，制造的；*v*.（大规模）制造（manufacture 的现在分词）；捏造；粗制滥造（文学作品）

food-manufacturing 食品生产

odors ['əudəz] *n*. 气味，名声（odor 的名词复数）

vapor ['veɪpə] *n*. 水汽，水蒸气，无实质之物；自夸者；幻想；[药] 吸入剂；[古] 忧郁（症）；*v*. 自夸，（使）蒸发

Exercises

Ⅰ. Answer the following questions according to the article

1. What is the definition of food sanitation?
2. What are the principles of food plant construction and design?
3. How to operate sanitary rules?
4. What are the main steps for executing sanitary controls?
5. How to wash hands in food plant?

Ⅱ. Choose a right term from what we have learnt to fill in each of the following blanks

1. Plant buildings and structures shall be suitable in ________, ________ and ________ to facilitate maintenance and sanitary operations for food-manufacturing purposes.

2. ________, ________, and ________ may be adequately cleaned and kept clean and kept in good repair.

3. Sanitary rules can be operated by using measures such as ________ maintenance and pest ________.

4. Single-service articles should be stored in ________ containers.

5. Hand-washing facilities shall be ________ and ________ and be furnished with ________ water at a suitable temperature.

Unit 8

Quality and Food Safety Assurance System

8.1 Introduction

Food Industry is one of the most important sectors in world economy with a highly significant relevance for economic and environmental development as well as social wellbeing. The food industry, as defined by the NACE is divided into sub-sectors: food and drink processing and manufacturing and food supply. The processing and manufacturing of food and drinks includes the following: meat, fish, fruit and vegetables, oils and fats, dairy, cereal related and starch products, beverages and sugar. The food supply includes wholesale and retail distribution of processed food and the catering sector. The food supply chain links a variety of activities: the procurement of agricultural raw materials, their processing up to final human consumption and their distribution. The food industry also involves multiple players such as farmers, input suppliers, manufacturers, packagers, transporters, exporters, wholesalers, retailers and final customers with different and changing interests, cultural attitudes and dimensions, which makes it a very dynamic and challenging industry.

Despite being specifically regulated, food industry has gone through many crises during the last decade: mad cow disease, Listeria, bird flu or the recent horsemeat scandal. The final consumer has become more and more sensitive to the origin and conservation of the products they buy. The logistics of concerned sector must be able to show responsiveness, accuracy and transparency to regain and maintain consumer confidence. The appearance of labels, continuous changes in international regulations as well as technological innovations have influenced and transformed the food supply and established principles like Product Traceability, Cold Chain Control or Hygiene and Quality. They involve the issue of transparency of the source of goods and merchandises as well as the delivery of raw materials for finished products which in turn is affected by flow management, particularly by the cold chain, either upstream or downstream of the transformation phase. Finally, their processing and storage are subject to very strict hygienic and quality conditions. Bar codes, electronic business standards, global data synchronization and Radio Frequency Identification (RFID) are some of the tools used to assure compliance with EU regulations and international standards. They supply a series of guidelines and standards on food safety in order to ensure fair practices in the international food trade, guarantee hygiene and provide healthy food products for consumption.

8.2 Risk and Controls in the Food Supply Chain

(1) Cold Chain

Hygienic safety of food depends largely on the respect of the cold chain, throughout all stages of storage and transport between producer, carrier, distributer and consumer. Throughout this chain, temperatures must not exceed the temperature regulations for each of the product categories, as are the negative cold products (quick and deep frozen food), ultra-fresh products (such as dairy products), fresh products (such as fruits and vegetables), chocolates or dry goods (such as groceries, beverages and liquor).

The cold chain can be broken:

• When products that are not to be kept at the same temperature are stored or transported together;

• Or when the food is too crowded, thus the cold does not penetrate into products depth;

• Or when the vehicle is not refrigerated in advance, the products can take in temperature until the truck is cooled or when loading or unloading is taking too long, there is a rapid loss of cold;

• When transporting or storing products that can be stored at different temperatures, the lowest temperature must be selected.

(2) Traceability

Traceability in the food industry is a key concern to all participants and stakeholders in the food chain. It refers to the ability to trace, through all stages of production, processing and distribution, the path of a food product, a food feed, a food-producing animal or a substance to be incorporated or even possibly incorporated into a food product or a food feed. All links involved, either professionals, producers, processors or distributors must identify and solve critical issues, maintain regulatory compliance, carry out their own self-controls while public service must establish and enforce regulations on hygiene control, consumers must be informed of nature of the products and know how to handle and store the products they buy through clearly identifiable labelling.

To improve the traceability, standards are implemented in the food sector on national and international basis. It aims to better control hazards and reduce risk levels. It is necessary for tracing the source of a problem of food poisoning or fraud.

The qualitative traceability is a device that combines both information relating to the physical flow of products and also to additional information on the product itself. This information can be, for example, the nature of its ingredients, the quantity, the origin of raw materials, the link between products, raw materials, finished product, etc.

Each intervenient within the food industry supply chain must identify:

• The goods received, processed and shipped (product type, producer's name and address);

• Suppliers and products delivered;

• Customers and products which are delivered to them;

• And record information related to goods as well as to be able to identify and recall a manufactured, processed or distributed product from the market.

Traceability can provide support to public health and help authorities determine the causes of contamination or help the companies reassure customers and increase competitiveness on the market through sales and market share.

8.3 Food Safety Assurance Systerm

Modern commerce in the food industry is both local and international in nature. Worldwide consumers are used to being able to obtain food products that are not locally grown or are not in season. They are critical of freshness, quality and safety. The development of and adherence to standards benefits everyone. Standards provide the framework for uniformity of quality and safety of products in commerce. They reduce cost and increase the efficiency of production. Internationally recognized standards open borders and markets for the free transport of foodstuffs around the world.

The importance of food safety standard may stand in the definition. Standards have been defined as parameters that segregate similar products into categories and describe them with consistent terminology that can be commonly understood by market participants. As standards improve the efficiency of markets, standards which concern any of the processes in the food chain.

Much of the current literature agreed that worldwide food industries applied Good Agricultural Practices (GAP), Hazard Analysis of Critical Control Points (HACCP) and International Organization for Standardization (ISO) as benchmark of food quality assurance. Those practices are in core triangle revolution of food quality system in food safety management. From the previous research, developing countries have been identified as lack of good agricultural, manufacturing and hygiene practices. Thus, the importance of total food safety management may boost their product for export market.

8.3.1 Principles and Systems for Quality and Food Safety Management

Hygiene, prevention and risk reduction, reliability, consistency, traceability, customer and consumer relevance, and transparency and accountability are the driving principles. They are operationalized through various management systems, some of them originating from the food industry, like HACCP, and some from other areas of industry, like 6 Sigma, quality function deployment and total productive maintenance. Certification schemes typically combine and package elements from various systems to fit the needs of a particular type of industry, and always represent a compromise between specificity and broad applicability. Moving forward, we may expect an ongoing drive to develop certification schemes around existing systems, to widen the applicability of existing schemes and to have the entire food supply chain covered by certification schemes.

8.3.2 5S Management System

5S is a simple tool for organizing your workplace in a clean, efficient and safe manner to enhance your productivity, visual management and to ensure the introduction of standardized working. 5S is a methodical way to organize your workplace and your working practices as well as being an overall philosophy and way of working. It is split into 5 phases, each named after a different Japanese term beginning with the letter "S"; (Seiri, Seiton, Seiso, Seiketsu, Shitsuke) hence the name 5S.

These five distinct phases are (with English descriptions):

Seiri: sort, clearing, classify;

Seiton: straighten, simplify, set in order, configure;

Seiso: sweep, shine, scrub, clean and check;

Seiketsu: standardize, stabilize, conformity;

Shitsuke: sustain, self-discipline, custom and practice.

Moreover, for completeness, some companies add a 6th (6S) of Safety, although in some opinion this should be an integral part of the steps of 5S and not a separate stage in itself.

8.3.3 GAP

Good Agricultural Practices (GAP) is a well-known public safety standard. GAP systems include a set of guideline for agricultural practices aiming at assuring minimum standards for production and storage. GAP systems underline pest management (optimal use of pesticides), manure handling at animal farms, maintenance of water quality, worker and field sanitation, guidelines for post-harvest handling and transportation, among others. Global GAP is a private safety standard that arisen from the GAPs. The existence of the private standard has been said to harmonize the public and private safety standard.

GLOBALG. A. P.'s roots began in 1997 as EUREPGAP, an initiative by retailers belonging to the Euro-Retailer Produce Working Group. British retailers working together with supermarkets in continental Europe become aware of consumers' growing concerns regarding product safety, environmental impact and the health, safety and welfare of workers and animals.

Harmonize their own standards and procedures and develop an independent certification system for Good Agricultural Practice (G. A. P.). The EUREPGAP standards helped producers comply with Europe-wide accepted criteria for food safety, sustainable production methods, worker and animal welfare, and responsible use of water, compound feed and plant propagation materials. Harmonized certification also meant savings for producers, as they would no longer need to undergo several audits against different criteria every year.

Over the next ten years, the process spread throughout the continent and beyond. Driven by the impacts of globalization, a growing number of producers and retailers around the globe joined in, gaining the European organization global significance. To reflect both its global reach and its goal of becoming the leading international G. A. P. standard, EUREPGAP changed its name to GLOBALG. A. P. in 2007. By 2008, the GLOBALGAP standard had expanded to cover coffee, tea, livestock, and aquaculture. Over 90000 producers in 87 countries had been certified.

GLOBALG. A. P. Certification covers:

• Food safety and traceability;

• Environment (including biodiversity);

• Workers' health, safety and welfare;

• Animal welfare;

• Includes Integrated Crop Management (ICM), Integrated Pest Control (IPC);

• Quality Management System (QMS), and Hazard Analysis and Critical Control Points (HACCP).

8.3.4 HACCP

HACCP is an internationally recognized system used to enhance food safety throughout the food chain. More and more companies around the world are using HACCP to prevent, reduce or eliminate potential food safety hazards, including those caused by cross-contamination.

The development of a HACCP system involves:

• Identifying potential hazards;

• Implementing control measures at specific points in the process;

• Monitoring and verifying that the control measures are working as intended.

Compared to traditional inspection procedures, HACCP:

• Provides a systematic approach to ensuring food safety;

• Gives more control over food safety to the processor;

• Is based on science, rather than simply past experience or subjective judgment;

• Focuses on preventing problems before they occur. This approach yields far better results than trying to detect failures through end-product testing.

There are two main elements of an effective HACCP system:

(1) Good Manufacturing Practices (GMP)

GMP are designed to control hazards related to plant personnel and the food-processing environment. Implementing GMP creates a safe and suitable environment for food processing. GMP include procedures and monitoring activities to help ensure that people and premises do not present food safety hazards. GMP lay the foundation for effective HACCP Plan, and must be developed and implemented prior to HACCP plans.

(2) HACCP plans

HACCP plans control hazards that are:

• Directly related to products, ingredients, and processes;

• Not covered by GMP;

• HACCP plans prevent, eliminate or reduce potential food safety hazards to an acceptable level, including hazards caused by cross-contamination.

Overview of steps to develop a HACCP plan:

• Describe your product, process, and hazards;

• Analyze your operations to identify any major food safety hazards;

• Put control measures in place at specific steps in the process to control major food safety hazards;

• Monitor how well the control measures work. If a hazard is not adequately controlled, take actions to correct the failure.

HACCP plans follow seven core principles. These principles were standardized by the Codex Alimentarius Commission (CAC). The Codex Alimentarius Commission was created by the Food and Agricultural Organization (FAO) and the World Health Organization (WHO) of the United Nations to develop food standards, guidelines, and related texts.

HACCP plans follow seven core principles:

(1) Conduct a Hazard Analysis

This involves:

- Identifying the hazards that might affect a particular product in a specific processing facility;
- Collecting and evaluating information on the hazards and the conditions leading to their presence;
- Deciding which hazards are significant to food safety, your operation must address these hazards through its HACCP plan(s).

(2) Determine the Critical Control Points

A critical control point (CCP) is a point, step or procedure in the process where food processors can apply a control measure. It is essential to prevent, eliminate or reduce a food safety hazard to an acceptable level. To determine the CCP in your process, you must identify where you can prevent, reduce or eliminate the hazards addressed in your HACCP plan.

(3) Establish Critical Limits

Critical limits are criteria that separate safe product from unsafe product. You must set critical limits for each CCP. Critical limits must be clearly defined and measurable.

(4) Establish Monitoring Procedures

Monitoring is the scheduled measurement or observation of a CCP relative to its critical limits. All monitoring results must be recorded.

(5) Establish Corrective Actions

If your monitoring detects a problem, there is a risk your operation has produced or will produce unsafe food. Your organization must have a plan in place to deal with these risks. For each CCP, you must document the corrective actions you will take to:

- Regain control of the hazard;
- Identify and control all affected product;
- Prevent the problem from happening again.

(6) Establish Verification Procedures

Verification procedures are used to determine if the HACCP system is working correctly. Verification involves methods, procedures, tests and other checks in addition to monitoring.

(7) Establish Record-Keeping and Documentation Procedures

You must document your HACCP plans, from start to finish. This includes all of the items listed above. All required monitoring and verification records must be complete and accurate. The principal of HACCP is to identify technological methods in the production process that can eliminate hazards (physical, chemical or biological). Apart from the critical control points, the producer has to establish limits of the critical control points, methods for their monitoring, and corrective steps. Verification of the whole system is an inseparable part of HACCP. The object of verification is to confirm that the system is working properly and the hazards are effectively kept under control5. For primary production where it is almost impossible to introduce HACCP but also for the food processing, GMP (Good Manufacturing Practice) is instigated. This is a system with the target of increasing food safety. It operates by setting rules for processing so that the risks of a harmful food occurrence are eliminated, and at the same time, the law is not broken. Principals of GMP are created for each production stage and describe precisely. The two systems, GMP and HACCP, follow up and are complementary. Besides the HACCP system, there are other systems, including BRC (British Retail Consortium), Food Safety System Certification 22000 or IFS (International Food Standard).

8.3.5 GMP

Good Manufacturing Practice (GMP) refers to advice and guidance put in place to outline the aspects of production and testing that can impact the quality and safety of a product. In the case of food and drink, GMP is aimed at ensuring that products are safe for the consumer and are consistently manufactured to a quality appropriate to their intended use. Manufacturers have for several years been driving towards such goals as Total Quality Management (TQM), lean manufacturing and sustainability-GMP is bound up with these issues. The ever-increasing interest amongst consumers, retailers and enforcement authorities in the conditions and practices in food manufacture and distribution, increases the need for the food manufacturer to operate within clearly defined policies such as those laid down in GMP. The ability to demonstrate that Good Manufacturing Practice has been fully and effectively implemented could, in the event of a consumer complaint or a legal action, reduce the manufacturer's liability and protect them from prosecution.

First launched in 1986, IFST's "Good Manufacturing Practice Guide" has been widely recognized as an indispensable reference work for food scientists and technologists. It sets out to ensure that food manufacturing processes deliver products that are uniform in quality, free from defects and contamination, and as safe as it is humanly possible to make them. This 6th edition has been completely revised and updated to include all the latest standards and guidance, especially with regard to legislation-driven areas such as HACCP.

The "Guide" is a must have for anyone in a managerial or technical capacity concerned with the manufacture, storage and distribution of food and drink. It is also a valuable reference for food education, training and for those involved in food safety and enforcement. Food scientists in academic and industry environments will value its precision, and policy makers and regulatory organizations will find it an indispensable guide to an important and multifaceted area.

8.3.6 ISO, the International Organization for Standardization

International Organization for Standardization (ISO) standard is an international standard in order to achieve uniformity and to prevent technical barriers to trade throughout the world. The essence of an ISO 9000-based quality system is that all activities and handling must be established in procedures, which must be followed by ensuring clear assignment of responsibilities and authorities. ISO International Standards provide practical tools for tackling many of today's global challenges, from managing global water resources to improving the safety of the food we eat. ISO International Standards create confidence in the products we eat or drink by ensuring the world uses the same recipe when it comes to quality, safety and efficiency. ISO has more than 1000 standards dedicated to food, covering everything from agricultural machinery to transportation, manufacturing and storage.

ISO (International Organization for Standardization) is the world's largest developer of voluntary International Standards providing benefits for business, government and society. ISO is a network comprising the national standards institutes of 163 countries. ISO standards make a positive contribution to the world we live in. They ensure vital features such as quality, ecology, safety, reliability, compatibility, interoperability, efficiency and effectiveness-and at an economical cost. They facilitate trade, spread knowledge, and share technological advances and good management practices. Today more than ever, food products regularly cross national boundaries at every stage of the supply chain, from farm to fork. ISO International Standards create confidence in the products we eat or drink.

ISO has developed a series of standards for food safety management systems that can be used by any organization in the food supply chain.

It features:

- ISO 22000:2005—Overall requirements (by the end of 2010, some 18630 certifications to ISO 22000 had been issued in 138 countries)
- ISO/TS 22002-1:2009—Specifi c prerequisites for food manufacturing
- ISO/TS 22002-3:2011—Specifi c prerequisites for farming
- ISO/TS 22003:2007—Guidelines for audit and certification bodies
- ISO 22004:2005—Guidelines for applying ISO 22000
- ISO 22005:2007—Traceability in the feed and food chain.

The ISO 22000 family contains a number of standards each focusing on different aspects of food safety management:

ISO 22000:2005 contains the overall guidelines for food safety management.

ISO 22004:2014 provides generic advice on the application of ISO 22000

ISO 22005:2007 focuses on traceability in the feed and food chain

ISO/TS 22002-1:2009 contains specific prerequisites for food manufacturing

ISO/TS 22002-2:2013 contains specific prerequisites for catering

ISO/TS 22002-3:2011 contains specific prerequisites for farming

ISO/TS 22002-4:2013 contains specific prerequisites for food packaging manufacturing

ISO/TS 22003:2013 provides guidelines for audit and certification bodies

8.4 Conclusion

The rapid pace of change in science and technology, changes in legislation and the current socioeconomic and sociodemographic realities have all had a marked impact on our food choices. Today, globalization makes it possible to have greater varieties of foods, brought to us from all corners of the world. As a result, food can now be sourced practically anywhere, sometimes subject to different quality standards and means of (pre-) preparation. This equates to additional risk and requires careful management at all levels across the food chain. Manufacturers and regulators alike have recognized their responsibilities, and are well aware just how vulnerable and unpredictable contamination can be if appropriate food safety measures are not firmly embedded in a manufacturer's food safety management system. Regaining the trust of consumers and developing an international consensus among stakeholders on the acceptable level of risks and the safety measures for effectively addressing these risks remains the key challenge for the 21st century. This chapter provides an overview of the modern approach to food safety management, roles of different sectors and the challenges and the outlook for the future. Developing countries need more support from Food Agriculture Organization (FAO) and World Health Organization (WHO) in terms of technical assistance, which may boost the development of food safety systems.

Technical Terms

NACE　Nomenclature of Economic Activities　经济活动命名
GAP　Good Agricultural Practices　良好农业规范
GLOBALG. A. P.　国际良好农业规范
HACCP　Hazard Analysis of Critical Control Points　危害分析与关键控制点
ISO　International Organization for Standardization　国际标准化组织
Euro-Retailer Produce Working Group　欧洲零售商协会
GMP　Good Manufacturing Practices　良好操作规范
CAC　Codex Alimentarius Commission　食品法典委员会
FAO　Food and Agricultural Organization　联合国粮食与农业组织
WHO　World Health Organization　世界卫生组织
BRC　British Retail Consortium　英国零售商协会
IFS　International Food Standard　国际食品标准
TQM　Total Quality Management　全面品质管理
IFST　Institute of Food Science and Technology　食品科学与技术学会
GMO　Genetically Modified Organisms　转基因生物
RFID　Radio Frequency Identification　无线射频识别
EU　European Union　欧洲联盟
Listeria [lɪ'stɪrɪə] *n*. 李斯特菌属
horsemeat ['hɔːsmiːt] *n*. 马肉
scandal ['skænd(ə)l] *n*. 丑闻；流言蜚语；诽谤；公愤
traceability [ˌtreɪsə'bɪlətɪ] *n*. [统计] 可追溯性；跟踪能力；可描绘
hygiene ['haɪdʒiːn] *n*. 卫生；卫生学；保健法
groceries ['grəʊsə] *n*. 杂货店；食品商
beverages [be'vəridʒiz] *n*. 饮料；酒水；饮料类（beverage的复数形式）
liquor ['lɪkə] *n*. 酒，含酒精饮料；溶液；液体；烈酒；*vt*. 使喝醉；*vi*. 喝酒，灌酒

certification [ˌsɜːtɪfɪˈkeɪʃn] *n*. 证明，保证；检定
compliance [kəmˈplaɪəns] *n*. 顺从，服从；承诺
segregate [ˈsegrɪgeɪt] *vt*. 使隔离；使分离；在……实行种族隔离
terminology [ˌtɜːmɪˈnɒlədʒɪ] *n*. 术语，术语学
harmonize [ˈhɑːmənaɪz] *vt*. 使和谐；使一致；以和声唱；*vi*. 协调；和谐；以和声唱

Exercises

Ⅰ. Answer the following questions according to the article

1. Why is the food industry to be said as a very dynamic and challenging industry?
2. Under what circumstances will the cold chain be broken?
3. Why should be standards implemented in the food sector on national and international basis?
4. What are principles for quality and food safety management?
5. Which 5 phases are involved in 5S Management System?
6. What dose the GAP focus on?
7. What is included in GLOBALG. A. P. certification?
8. What is the feature of HACCP compared to traditional inspection procedures?
9. When you establish the corrective actions for each CCP, what should be noticed?
10. What is the core principle of HACCP plan?
11. What is the definition of GMP?
12. What dose GMP contain?
13. What is included in ISO?

Ⅱ. Choose a term from what we have learnt to fill in each of the following blanks. Change the word form where necessary

1. Food industry as defined by the NACE is divided into sub-sectors: ________ and ________.
2. The processing and manufacturing of food and drinks includes: ____________ ____________________.
3. The food supply includes ____________________ and ____________________.
4. EU supply a series of guidelines and standards on food safety in order to ensure ________, ________ and provide ________ for consumption.
5. Traceability can provide ________ and help ________ or help ________ and increase ________ on the market through sales and market share.
6. Worldwide food industries applied ________, ________ and ________ as benchmark of food quality assurance.
7. 5S is a simple tool for organizing your workplace in a ________, ________ and ________ manner to enhance your ________, ________ and to ensure ________.
8. GAP systems include a set of guideline for agricultural practices aiming at ________ ________________.
9. More and more companies around the world are using HACCP to ________, ________ or ________ potential food safety hazards, including those caused by cross-contami-

nation.

10. HACCP plans follow ________ core principles. These principles were standardized by the ________. The CAC was created by the ________ and the ________ to develop food standards, guidelines, and related texts.

11. A critical control point (CCP) is a ________, ________ or ________ in the process where food processors can apply a control measure.

12. A GMP is a system for ensuring that products are consistently produced and controlled according to ________.

13. International Organization for Standardization (ISO) standard is an international standard in order to achieve ________ and to prevent ________ to trade throughout the world.

14. The essence of an ________ is that all activities and handling must be established in procedures, which must be followed by ensuring clear assignment of responsibilities and authorities.

15. Look ahead, developing countries need more support from ________ and ________ in terms of technical assistance.

Unit 9

Food Safety and Quality International Organization—CAC

Since ancient times, a series of food standards and laws have been enacted by authorities to protect the consumers against unsafe, adulterated, and misbranded food. With technological advancements in storage and transportation, the twentieth century has witnessed an exponential increase in food trade. However, the conflicting or contradictory nature of regulations in different countries led to barriers in trade and affected the food distribution, a harmonized layout for food commodity standards were imperative to facilitate the movement of food across borders.

Food and Agriculture Organization (FAO) and World Health Organization (WHO) thus began a series of joint expert meetings to harmonize the food commodity standards worldwide. The Joint FAO/WHO food standards conference, convened in Geneva in 1962, established the framework for cooperation between the two agencies. The Codex Alimentarius Commission (CAC) was endorsed in 1963 at the sixteenth World Health Assembly and to be the body responsible for implementing the Joint FAO/WHO food standards programme. In the same year, the first session of the CAC was held in Rome, and it was attended by delegates from 30 countries and 16 international organizations. Currently, the CAC has 188 Codex Members, 187 member countries, and 1 member organization (European Union, EN); 240 Codex Observers, 56 international governmental organizations, 168 nongovernmental organizations, and 16 United Nations (UN) agencies. The membership to the CAC is open to all member nations and associate members of the FAO and/or WHO. The headquarters of the CAC is based in Rome and the meetings of CAC are held in every alternate year in either Rome or Geneva. Representation at sessions is on a country basis. Special or extraordinary sessions can be scheduled if needed.

9.1 Purpose of CAC

The Codex Alimentarius is a collection of internationally adopted food standards, guidelines, and codes of practice, covering food safety matters (residues, hygiene, additives, contaminants, etc.) and quality matters (product descriptions, quality classes, labeling, and certification), which serves as the basis for many national food standards and related regulations. The CAC is always committed to develop internationally agreed standards and related texts for use in domestic regulation and international trade in food that are based on scientific principles and fulfill the objectives of consumer health protection and fair practices in food trade. Another of the principal purposes of the CAC is the preparation of food standards and the publication of the Codex Alimentarius, which intended to guide and to promote the elaboration and establishment of definitions and requirements for foods to assist in their harmonization. The CAC also promotes the coordination of all food standards work undertaken by international governmental and nongovernmental organizations, such as World Organization for Animal Health (OIE) and International Plant Protection Convention (IPPC).

Furthermore, the CAC and its specialized technical subsidiary bodies provide structured, neutral meetings for discussion of all topics related to food safety and trade with in its mandate. Representatives from governments, consumer groups, industry, and academia meet to exchange views about food safety and trade and to adopt standards or related texts. Countries that are not yet members of the CAC sometimes attend in an observer capacity.

9.2 Scope of the Codex Alimentarius

The Codex Alimentarius includes standards/guidelines/codes of practices for food hygiene, food additives, residues of pesticides and veterinary drugs, contaminants, commodities (e. g., milk, meat, fruits and vegetables, and processed food), labeling and presentation, methods of analysis and sampling, and import and export inspection and certification. Thus, it looks at both horizontal and vertical standard setting as far as food is concerned.

9.3 Structure and Subsidiary Bodies Under the CAC

The CAC consists of the four main organizational elements as follows, (a) The Commission; (b) The Executive Committee; (c) The Codex Subsidiary bodies; (d) The Codex Secretariat, each of which has a positive role to play in the achieving the mandate of the CAC.

The Executive Committee acts on behalf of the CAC as its executive organ between sessions of the Commission. In particular, the Executive Committee may make proposals to the CAC regarding general orientation, strategic planning, and programming of the work; study special problems; and assists in the management of the Commission’s program of standards development, namely by conducting a critical review of proposals to undertake work and by monitoring the progress of standards development.

The Codex Committees, the Joint FAO/WHO Regional Coordinating Committees and ad hoc Intergovernmental Task Force (TF) are three kinds of subsidiary bodies under the CAC. The Codex Committees, further divided into the General Subject Committees and the Commodity Committees, prepare draft standards for submission to the CAC or deal with Codex procedures. General Subject Committees work on issues that can apply to any commodity or groups of commodities. They develop concepts and principles applying to foods in general, specific foods, or groups of foods, and endorse or review relevant provisions in Codex commodity standards. Based on the advice of expert scientific bodies, they develop major recommendations pertaining to consumers' health and safety. The responsibility for developing standards for specific foods or classes of food lies with the Commodity Committees. Commodity Committees convene as necessary and go into recess or are abolished when the Commission decides their work has been completed. The Coordinating Committees is responsible for the coordination of food standards activities and the development of regional standard, ensuring that the CAC work is responsive to regional interests and to the concerns of developing countries. Coordinating Committees are usually hosted in the country of the coordinator (but may be cohosted in other countries) and the main costs (interpretation and translation) are funded by the Codex budget that is covered entirely by FAO and WHO. Ad hoc Intergovernmental Task Force (TF) is established for one specific purpose with very limited terms of reference (TOR) and are dissolved when their task is completed.

The Codex Secretariat, located at FAO headquarters in Rome, organizes the meetings of the CAC and the Executive Committee, and it facilitates the work of the subsidiary bodies in close coordination with the secretariat of the host country. The Codex Secretariat is a team of less than 20 professional and technical staff, deals with all aspects of the functioning of the CAC and its subsidiary bodies (e. g., preparing agendas, distributing working documents, drafting the reports of the sessions) and with the updating of the Codex Alimentarius. Organizational structure of the Codex Alimentarius Commission see Fig 9. 1.

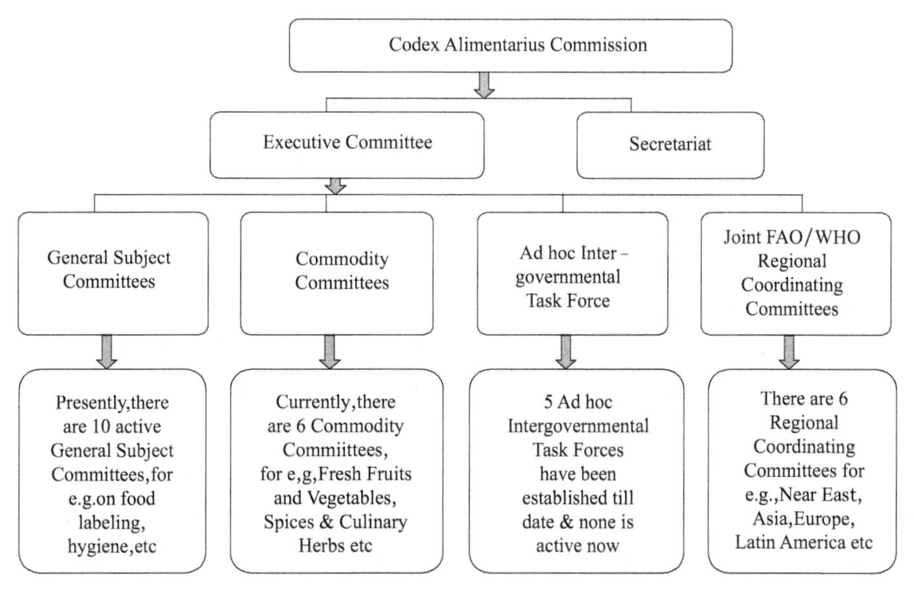

Fig 9. 1 Organizational structure of the Codex Alimentarius Commission (CAC)

9.4 How the CAC Works

The CAC is guided in its work by a strategic plan, which is reviewed every 5 years. The plan contains the following five goals that have been broken down further into a number of activities:

- Goal 1: Promoting sound regulatory frameworks;

• Goal 2: Promoting widest and consistent application of scientific principles and risk analysis;

• Goal 3: Strengthening codex work-management capabilities;

• Goal 4: Promoting cooperation between codex and relevant international organizations;

• Goal 5: Promoting maximum and effective participation of members.

9.5 Role of Codex Standards in the Framework of International Trade

The Codex Alimentarius serves as the basis for many national food standards and related regulations, which have become international references for food safety under the WTO' s agreement. This means that, as long as WTO members apply CAC standards for food safety measures, no other member can challenge such measures as unjustified barriers to international trade. If a country wants to set stricter standards, this needs to be justified scientifically through a risk analysis process. In case of a dispute, Codex standards and related texts are also recognized as international standards on Technical Barriers to Trade (TBT agreement) and used in relevant dispute settlements procedures. The Codex standards have assumed considerable significance under the technical regulations and standards provisions contained in the TBT agreement. The recognition of the scientifically based Codex standards and the related texts within the framework of the TBT agreements has given a big flip to the work of Codex and generated more interest in its members. There has been increased member attendance at Codex meetings, particularly from the developing countries.

9.6 The Codex Alimentarius Today and in the Future

Today consumers enjoy a variety of food from all over the world. The Codex Alimentarius rules for food labeling, food additives, pesticide residues, contaminants, food hygiene, and others provide a foundation for ensuring food safety and nutrient content of food. Therefore, consumers can be more confident about the safety and quality of the food they consume, regardless of its origin. Meanwhile, the global market for trade in food continues to expand. More and more countries participate actively in the standard-setting of the CAC and have adopted the standards on food production and processing, thereby facilitating food trade and contributing to the economic health of countries and regions around the world. Once producers meet these standards, they can be confident that their products are safe, of high quality, and are acceptable in export markets. Hence, more and more countries have revise their food laws and regulations in accordance with Codex Alimentarius standards and strengthening their national food control systems to improve food quality and safety.

Despite the importance of Codex in protecting health and facilitating trade, many countries have not been able to participate fully in the establishment of international food safety standards in Codex, mainly because of low-income levels. To redress this situation, the Directors-General of FAO and WHO launched the Codex Trust Fund in 2003, which is aimed at assisting developing countries and transition economy countries to enhance their level of effective participation in the CAC. This is a vital step toward ensuring that the Codex system is inclusive, participatory, and equitable. Hence, the keys to the success of Codex are that it is member driven, science based, transparent, inclusive, flexible, and based on consensus. This allows it to take on any new challenging issue brought up by a member state. At the same time, the Codex also faced with the challenges today. There is a need to make the process of standard setting more inclusive by making an extra effort to encourage the developing countries to submit the necessary data and participate in a transaction of the CAC. The consensus among members with different socioeconomic status and trade interest/needs was required to find, the use of private standards that may have different/stricter requirements than Codex standards was required to increase. The CAC must be more nimble-footed in setting standards to maintain relevance, equity, inclusive and transparent in the rapidly changing environment.

In conclusion, the CAC reinforces the extremely important responsibility of laying down the global parameters for the quality and safety of food products for human consumption that the CAC is shouldering.

Technical Terms

Codex Alimentarius Commission (CAC) 食品法典委员会
enact [ɪ'nækt] *vt.* 制定法律，规定，颁布，担任……角色
adulterated [ə'dʌltəreɪtɪd] *adj.* 掺入次级品的
misbranded [mɪs'brænd] *adj.* 虚假标记的
exponential [ˌekspə'nenʃl] *adj.* 指数的，幂数的，越来越快的
imperative [ɪm'perətɪv] *adj.* 势在必行的，必要的，不可避免的
harmonize ['hɑːmənaɪz] *vt.* 使和谐，为（旋律）配和声
endorse [ɪn'dɔːs] *vt.* 签署，批准，签署（签名），支持，核准
hygiene ['haɪdʒiːn] *n.* 卫生，卫生学
certification [ˌsɜːtɪfɪ'keɪʃn] *n.* 证明，鉴定，证书
mandate ['mændeɪt] *n.* 授权，命令，委任，任期
inspection [ɪn'spekʃn] *n.* 检验，检查
pertaining [pə(ː)'teɪnɪŋ] *adj.* 与……有关系的，附属……的，为……固有的（to）
redress [rɪ'dres] *vt.* 补偿，改正，矫正，革除，重新调整
inclusive [ɪn'kluːsɪv] *adj.* 包括的，包罗广泛的
transaction [træn'zækʃn] *n.* 业务，事务

Exercises

Ⅰ. Answer the following questions according to the article

1. What is the purpose of CAC?
2. What are included within the scopes of CAC standards?
3. What are the organizational structures and operating mechanism of CAC?
4. What does CAC standards play role in the framework of international trade?
5. How to evaluate the future and challenges?

Ⅱ. Choose a term from what we have learnt to fill in each of the following blanks. Change the word form where necessary.

1. The subsidiary bodies of CAC are ________, ________, ________.

2. The Codex Alimentarius includes standards/guidelines/codes of practices for ________, ________, ________, ________, ________, ________, and so on.

3. The Codex Alimentarius develops harmonized international food standards, guidelines, and codes of practice to ________________, and to ________________.

4. The Codex Alimentarius is a collection of ________________ and related texts presented in a uniform manner.

5. Codex standards and related texts are also recognized as ________ on TBT agreement and used in ________________.

Unit 10

Introduction of cGMP, SSOP, HACCP, ISO 9000 and ISO 22000

10.1 Introduction

Food safety has become a constant concern all over the world, leading healthcare institutions and governments of several countries to find ways to monitor production chains. In this context, it is essential that quality management tools are adopted. These tools should emphasize the standardization of products and process, product traceability, and food-safety assurance. The basis of the food-safety system to be adopted in the food industry consists of a combination of good manufacturing practices (GMP), sanitation standard operating procedures (SSOP), and a hazard analysis and critical control point (HACCP) system.

10.2 Prerequisite Programs

10.2.1 General Principles and Definitions

GMP and SSOP are prerequisite programs for HACCP implementation. Prerequisite programs deal with the "good housekeeping" issues in the facility and may prevent a hazard from occurring, whereas HACCP manages specific hazards within the process. The plant must provide all documentation including programs in written records and results for all prerequisite programs that support their HACCP system. It is important that the management of the company be committed with the objectives of the system plan and collaborate with the multidisciplinary team responsible for creating and implementing the plan. The HACCP team should evaluate prerequisite programs at the moment of inspection and determine if these programs continue to support the decision in the hazard analysis step of HACCP system implementation.

10.2.2 Current Food Good Manufacturing Practices (cGMP)

Current Good Manufacturing Processes (cGMP), also known as good manufacturing Processes (GMP), provide guidelines for manufacturing, testing, and quality assurance to ensure that a product is safe for human or animal consumption or use.

The current cGMP consist of seven subparts, two of which are reserved (Subpart D & F) (Table 10.1). The requirements are purposely general to allow individual variation by manufacturers to implement the requirements in a manner that best suit their needs.

Table 10.1 Summary of 21 CFR Part 110: Current Good Manufacturing Practice in Manufacturing, Packing, or Holding Human Food

Subpart A. General Provisions	Section 110.3	Definitions	Definitions of: • Acid foods/acidified foods • Adequate • Batter • Blanching • Critical control point • Food • Food-contact surfaces • Lot • Microorganisms • Pest • Plant • Quality control operation • Rework • Safe-moisture level • Sanitize • Shall • Should • Water activity
	Section 110.5	Current good manufacturing practice	• Criteria for determining adulteration • Food covered by specific GMP is also covered by umbrella GMP
	Section 110.10	Personnel	Requirements for: • Disease control • Cleanliness • Education and training • Supervision of personnel with regards to these requirements
	Section 110.19	Exclusions	• Excluded operations (raw agricultural commodities) • FDA can issue special regulations to cover excluded operations
Subpart B. Buildings and Facilities	Section 110.20	Plant and Grounds	• Description of adequate maintenance of grounds • Plant construction and design to facilitate sanitary operations and maintenance
	Section 110.35	Sanitary Operations	Requirements for: • Cleaning/sanitizing of physical facilities, utensils, and equipment • Storage of cleaning and sanitizing substances • Pest control • Sanitation of food contact surfaces • Storage and handling of cleaned portable equipment and utensils
	Section 110.37	Sanitary Facilities and Controls	Requirements for: • Water supply • Plumbing • Sewage disposal • Toilet facilities • Hand-washing facilities • Rubbish and offal disposal

续表

Subpart C.Equipment	Section 110. 40	Equipment and Utensils	• Requirements for the design, construction, and maintenance of equipment and utensils
Subpart E. Production and Process Controls	Section 110. 80	Processes and controls	Delineates processes and controls for: • Raw materials and other ingredients • Manufacturing operations
Subpart G.Defect Action Levels	Section 110. 10		• FDA has established maximum defect action levels (DALs) for some natural or unavoidable defects • Compliance with DALs does not excuse violation of 402 (a)(4) • Food containing defects above DALs may not be mixed with other foods

Source: Federal Register 51, 1986.

10. 2. 3 Sanitation Standard Operating Procedures (SSOP)

Sanitation standard operating procedures (SSOP) refer to sanitary action taken during production to prevent product contamination or adulteration. These practices or procedures must be documented daily to validate that food product safety was maintained during production. SSOP should be specific to each facility where almonds or almond products are produced. Develop your own forms, or use those provided at the end of this section.

Plant sanitation must address facility environment, processing equipment, procession equipment, and all employees. A designated sanitarian that has satisfactorily completed a certified food sanitation program should be in charge of writing formal plans and procedures. The sanitarian and line supervisors should visually inspect processing equipment to ensure that proper sanitation has been completed. The SSOP should specify how the handler will meet than sanitation conditions and practices that are to be monitored.

(1) Sanitation Controls

Each processor shall have and implement a sanitation standard operating procedure (SSOP) that addresses sanitation conditions and practices before, during, and after processing. The SSOP shall address:

• Safety of the water that comes into contact with food or food contact surfaces or that is used in the manufacture of ice;

• Condition and cleanliness of food contact surfaces, including utensils, gloves, and outer garments;

• Prevention of cross contamination from insanitary objects to food, food packaging material, and other food contact surfaces, including utensils, gloves, and outer garments, and from raw product to processed product;

• Maintenance of hand washing, hand sanitizing, and toilet facilities;

• Protection of food, food packaging material, and food contact surfaces from adulteration with lubricants, fuel, pesticides, cleaning compounds, sanitizing agents, condensate, and other chemical, physical, and biological contaminants;

• Proper labeling, storage, and use of toxic compounds;

• Control of employee health conditions that could result in the microbiological contamination of food, food packaging materials, and food contact surfaces; and

• Exclusion of pests from the food plant.

(2) Monitoring

The processor shall monitor the conditions and practices during processing with sufficient frequency to ensure, at a minimum, conformance with those conditions and practices specified in part 110 of this chapter and in subpart B of part 117 of this chapter that are appropriate both to the plant and to the food being processed. Each processor shall correct, in a timely manner, those conditions and practices that are not met.

(3) Records

Each processor shall maintain SSOP records that, at a minimum, document the monitoring and corrections prescribed. These records are subject to the recordkeeping requirements of 120.12.

10.3 Hazard Analysis and Critical Control Points(HACCP)

The Hazard Analysis and Critical Control Points (HACCP) system, which is science based and systematic, identifies specific hazards and measures for their control to ensure the safety of food. HACCP is a tool to assess hazards and establish control systems that focus on prevention rather than relying mainly on end-product testing. Any HACCP system is capable of accommodating change, such as advances in equipment design, processing procedures or technological developments.

HACCP can be applied throughout the food chain from primary production to final consumption and its implementation should be guided by scientific evidence of risks to human health. As well as enhancing food safety, implementation of HACCP can provide other significant benefits. In addition, the application of HACCP systems can aid inspection by regulatory authorities and promote international trade by increasing confidence in food safety.

The successful application of HACCP requires the full commitment and involvement of management and the work force. It also requires a multidisciplinary approach; this multidisciplinary approach should include, when appropriate, expertise in agronomy, veterinary health, production, microbiology, medicine, public health, food technology, environmental health, chemistry and engineering, according to the particular study. The application of HACCP is compatible with the implementation of quality management systems, such as the ISO 9000 series, and is the system of choice in the management of food safety within such systems.

While the application of HACCP to food safety was considered here, the concept can be applied to other aspects of food quality.

Developing a HACCP plan

In the development of a HACCP plan, five preliminary tasks need to be accomplished before the application of the HACCP principles to a specific product and process.

① Step 1 Assemble the HACCP Team

The food operation should assure that the appropriate product specific knowledge and expertise is available for the development of an effective HACCP plan. Optimally, this may be accomplished by assembling a multidisciplinary team. Where such expertise is not available on site, expert advice should be obtained from other sources. The scope of the HACCP plan should be identified. The scope should describe which segment of the food chain is involved and the general classes of hazards to be addressed (e. g. does it cover all classes of hazards or only selected classes).

② Step 2 Describe the food and its distribution

A full description of the product should be drawn up, including relevant safety information such as composition, physical/chemical structure (including A_w, pH, etc.), microcidal/static treatments (heat-treatment, freezing, brining, smoking, etc.), packaging, durability and storage conditions and method of distribution.

③ Step 3 Identify intended use

The intended use should be based on the expected uses of the product by the end user or consumer. In specific cases, vulnerable groups of the population, e. g. institutional feeding, may have to be considered.

④ Step 4 Construct flow diagram

The flow diagram should be constructed by the HACCP team. The flow diagram should cover all steps in the operation. When applying HACCP to a given operation, consideration should be given to steps preceding and following the specified operation.

⑤ Step 5 On-site confirmation of flow diagram

The HACCP team should confirm the processing operation against the flow diagram during all stages and hours of operation and amend the flow diagram where appropriate.

⑥ Step 6 List all potential hazards associated with each step, conduct a hazard analysis, and consider any measures to control identified hazards (Principle 1)

The HACCP team should list all of the hazards that may be reasonably expected to occur at each step from primary production, processing, manufacture, and distribution until the point of consumption. The HACCP team should next conduct a hazard analysis to identify for the HACCP plan which hazards are of such a nature that their elimination or reduction to acceptable levels is essential to the production of a safe food.

In conducting the hazard analysis, wherever possible the following should be included:

- the likely occurrence of hazards and severity of their adverse health effects;
- the qualitative and/or quantitative evaluation of the presence of hazards;
- survival or multiplication of microorganisms of concern;
- production or persistence in foods of toxins, chemicals or physical agents; and,
- conditions leading to the above.

The HACCP team must then consider what control measures, if any, exist which can be applied for each hazard. More than one control measure may be required to control a specific hazard(s) and more than one hazard may be controlled by a specified control measure.

⑦ Step 7 Determine critical control points (CCP) (Principle 2)

There may be more than one CCP at which control is applied to address the same hazard. The determination of a CCP in the HACCP system can be facilitated by the application of a decision tree, which indicates a logic reasoning approach. Application of a decision tree should be flexible, given whether the operation is for production, slaughter, processing, storage, distribution or other. It should be used for guidance when determining CCP.

If a hazard has been identified at a step where control is necessary for safety, and no control measure exists at that step, or any other, then the product or process should be modified at that step, or at any earlier or later stage, to include a control measure.

⑧ Step 8 Establish critical limits (Principle 3)

Critical limits must be specified and validated if possible for each CCP. In some cases more than one critical limit will be elaborated at a particular step. Criteria often used include measurements of temperature, time, moisture level, pH, A_w, available chlorine, and sensory parameters such as visual appearance and texture.

⑨ Step 9 Establish monitoring procedures (Principle 4)

Monitoring is the scheduled measurement or observation of a CCP relative to its critical limits. The monitoring procedures must be able to detect loss of control at the CCP. Further, monitoring should ideally provide this information in time to make adjustments to ensure control of the process to prevent violating the critical limits. Where possible, process adjustments should be made when monitoring results indicate a trend towards loss of control at a CCP. The adjustments should be taken before a deviation occurs. Data derived from monitoring must be evaluated by a designated person who has got knowledge and authority to carry out corrective actions when indicated. If monitoring is not continuous, then the amount or frequency of monitoring must be sufficient to guarantee the CCP is in control. Most monitoring procedures for CCP will need to be done rapidly because they relate to online processes and there will not be time for lengthy analytical testing. Physical and chemical measurements are often preferred to microbiological testing because they may be done rapidly and can often indicate the microbiological control of the product. All records and documents associated with monitoring CCP must be signed by the person(s) doing the monitoring and by a responsible reviewing official(s) of the company.

⑩ Step 10 Establish corrective action (Principle 5)

Specific corrective actions must be developed for each CCP in the HACCP system in order to deal with deviations when they occur.

The actions must ensure that the CCP has been brought under control. Actions taken must also include proper disposition of the affected product. Deviation and product disposition procedures must be documented in the HACCP record keeping.

⑪ Step 11 Establish verification procedures (Principle 6)

Establish procedures for verification. Verification and auditing methods, procedures and tests, including random sampling and analysis, can be used to determine if the HACCP system is working correctly. The frequency of verification should be sufficient to confirm that the HACCP system is working effectively. Examples of verification activities include:

- Review of the HACCP system and its records;
- Review of deviations and product dispositions;
- Confirmation that CCP are kept under control.

Where possible, validation activities should include actions to confirm the efficacy of all elements of the HACCP plan.

⑫ Step 12 Establish record-keeping and documentation procedures (Principle 7)

Efficient and accurate record keeping is essential to the application of a HACCP system. HACCP procedures should be documented. Documentation and record keeping should be appropriate to the nature and size of the operation.

Documentation examples are:

• Hazard analysis;

• CCP determination;

• Critical limit determination.

Record examples are:

• CCP monitoring activities;

• Deviations and associated corrective actions;

• Modifications to the HACCP system.

10.4 ISO 9000—Quality Management

ISO 9000 is a set of international standards on quality management and quality assurance developed to help companies effectively document the quality system elements to be implemented to maintain an efficient quality system.

The ISO 9000 family addresses various aspects of quality management and contains some of ISO's best-known standards. The standards provide guidance and tools for companies and organizations who want to ensure that their products and services consistently meet customer's requirements, and that quality is consistently improved.

Standards in the ISO 9000 family include:

• ISO 9001:2015 Quality management systems—Requirements

• ISO 9000:2015 Quality management systems—Fundamentals and Vocabulary (definitions)

• ISO 9004:2009 Quality management systems-Managing for the sustained success of an organization (continuous improvement)

• ISO 19011:2011 Guidelines for auditing management systems

10.4.1 ISO 9001: 2015 Quality Management Systems—Requirements

ISO 9000:2015 specifies requirements for a quality management system when an organization:

• Needs to demonstrate its ability to consistently provide products and services that meet customer and applicable statutory and regulatory requirements, and

• Aims to enhance customer satisfaction through the effective application of the system, including processes for improvement of the system and the assurance of conformity to customer and applicable statutory and regulatory requirements.

10.4.2 ISO 9000:2015 Quality Management Systems—Fundamentals and Vocabulary (definitions)

ISO 9000:2015 describes the fundamental concepts and principles of quality management that are universally applicable to the following:

• Organizations seeking sustained success through the implementation of a quality management system;

• Customers seeking confidence in an organization's ability to consistently provide products and services conforming to their requirements;

• Organizations seeking confidence in their supply chain that their product and service requirements will be met;

• Organizations and interested parties seeking to improve communication through a common understanding of the vocabulary used in quality management;

• Organizations performing conformity assessments against the requirements of ISO 9001;

• Providers of training, assessment or advice in quality management;

• Developers of related standards.

ISO 9000:2015 specifies the terms and definitions that apply to all quality management and quality management system standards developed by ISO/TC 176.

10.4.3 ISO 9004:2009 Quality Management Systems—Managing for the Sustained Success of an Organization (continuous improvement)

ISO 9004:2009 provides guidance to organizations to support the achievement of sustained success by a quality management approach. It is applicable to any organization, regardless of size, type and activity.

ISO 9004:2009 is not intended for certification, regulatory or contractual use.

10.4.4 ISO 19011:2011 Guidelines for Auditing Management Systems

ISO 19011:2011 provides guidance on auditing management systems, including the principles of auditing, managing an audit programme and conducting management system audits, as well as guidance on the evaluation of competence of individuals involved in the audit process, including the person managing the audit programme, auditors and audit teams.

ISO 19011: 2011 is applicable to all organizations that need to conduct internal or external audits of management systems or manage an audit programme.

The application of ISO 19011:2011 to other types of audits is possible, provided that special consideration is given to the specific competence needed.

10.5 ISO 22000:2005 Food and Safety Management Systerms-Requirement for Organization in the Food Chain

ISO 22000:2005 is an international standard and defines the requirements of a food safety management system covering all organizations in the food chain from "farm to fork", including catering and packaging companies.

The standard combines generally recognized key elements to ensure food safety along the food chain including: interactive communication; system management; control of food safety hazards through pre-requisite programmes and HACCP plans; and continual improvement and updating of the management system.

ISO 22000:2005 family include the following:

- ISO 22000 Food safety management systems—Requirements for organizations throughout the food chain;
- ISO 22003 Food safety management—Requirements for bodies providing audit and certification of food safety management systems;
- ISO 22004 Food safety management—Guidance on the application of ISO 22000: 2005;
- ISO 22005 Traceability in the feed and food chain—General principles and basic requirements for system design and implementation;
- ISO 22006 Quality management systems—Guidance on the application of ISO 9001: 2000 for crop production data;
- ISO 2200X Quality management systems—Basic hygiene elements for prerequisite programmes in food producing and handling organization.

This International Standard specifies the requirements for a food safety management system that combines the following generally recognized key elements to ensure food safety along the food chain, up to the point of final consumption:

- Interactive communication;
- System management;
- Prerequisite programmes;
- HACCP principles.

Communication along the food chain is essential to ensure that all relevant food safety hazards are identified and adequately controlled at each step within the food chain. This implies communication between organizations both upstream and downstream in the food chain. Communication with customers and suppliers about identified hazards and control measures will assist in clarifying customer and supplier requirements (e. g. with regard to the feasibility and need for these requirements and their impact on the end product).

Recognition of the organization's role and position within the food chain is essential to ensure effective interactive communication throughout the chain in order to deliver safe food products to the final consumer. An example of the communication channels among interested parties of the food chain is shown in Fig. 10. 1.

The most effective food safety systems are established, operated and updated within the framework of a structured management system and incorporated into the overall management activities of the organization. This provides maximum benefit for the organization and interested parties. This International Standard has been aligned with ISO 9001 in order to enhance the compatibility of the two standards.

This International Standard can be applied independently of other management system standards. Its implementation can be aligned or integrated with existing related management system requirements, while organizations may utilize existing management system(s) to establish a food safety management system that complies with the requirements of this International Standard.

This International Standard integrates the principles of the Hazard Analysis and Critical Control Point (HACCP) system and application steps developed by the Codex Alimentarius

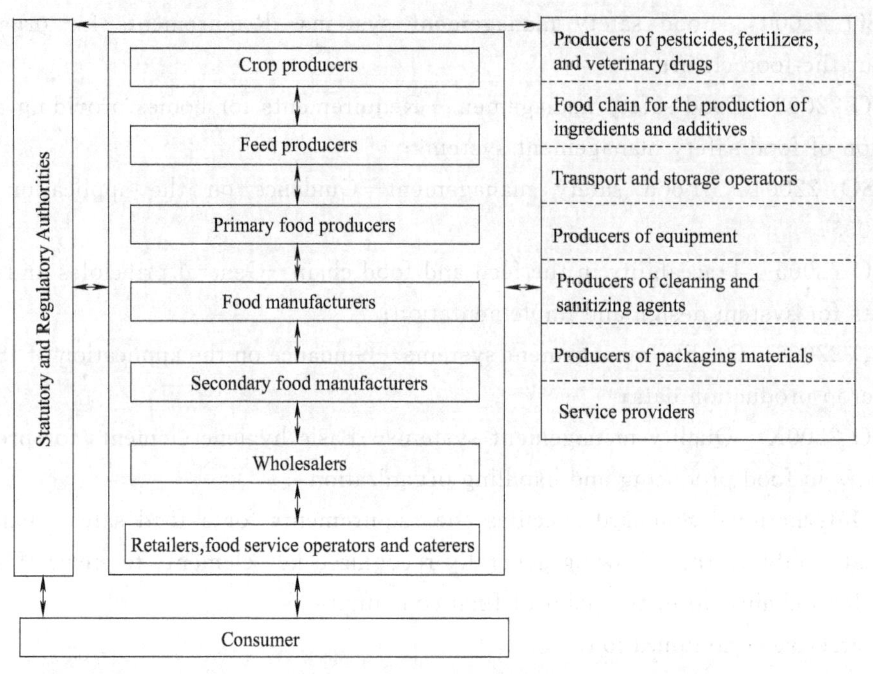

Figure 10.1 Example of communication within the food chain

Commission. By means of auditable requirements, it combines the HACCP plan with prerequisite programmes (PRP). Hazard analysis is the key to an effective food safety management system, since conducting a hazard analysis assists in organizing the knowledge required to establish an effective combination of control measures. This International Standard requires that all hazards that may be reasonably expected to occur in the food chain, including hazards that may be associated with the type of process and facilities used, are identified and assessed. Thus, it provides the means to determine and document why certain identified hazard need to be controlled by a particular organization and why others need not.

During hazard analysis, the organization determines the strategy to be used to ensure hazard control by combining the PRP(s), operational PRP(s) and the HACCP plan.

To facilitate the application of this International Standard, it has been developed as an auditable standard. However, individual organizations are free to choose the necessary methods and approaches to fulfil the requirements of this International Standard. To assist individual organizations with the implementation of this International Standard, guidance on its use is provided in ISO/TS 22004.

This International Standard is intended to address aspects of food safety concerns only. The same approach as provided by this International Standard can be used to organize and respond to other food specific aspects (e. g. ethical issues and consumer awareness).

This International Standard allows an organization (such as a small and/or less developed organization) to implement an externally developed combination of control measures.

The aim of this International Standard is to harmonize on a global level the requirements for food safety management for businesses within the food chain. It is particularly intended for application by organizations that seek a more focused, coherent and integrated food safety management system than is normally required by law. It requires an organization to meet any applicable food safety related statutory and regulatory requirements through its food safety management system.

Technical Terms

traceability [ˌtreɪsəˈbɪlətɪ] *n*. 可追溯，追源，可追踪性

Good Manufacturing Practices (GMP) 良好操作规范

Sanitation Standard Operating Procedures (SSOP) 卫生标准操作程序

Hazard Analysis and Critical Control Point (HACCP) 危害分析与关键控制点

prerequisite [priːˈrekwəzɪt] *n*. 先决条件，前提；*adj*. 必须先具备的，必要的，先决条件的

prerequisite programs 前提方案

implementation [ˌɪmplɪmenˈteɪʃn] *n*. 成就，贯彻，按照启用

multidisciplinary [ˌmʌltidɪsəˈplɪnəri] *adj*. 多学科，多部门，多科目

plumbing [ˈplʌmɪŋ] *n*. 水管装置，管道工程

delineate [dɪˈlɪnieɪt] *vt*. 勾画，描述

Food and Drug Administration (FDA) 食品和药品管理局

Defect Action Levels (DALs) 缺陷行动水平

adulteration [əˌdʌltəˈreɪʃn] *n*. 掺假，掺杂

validate [ˈvælɪdeɪt] *vt*. 使合法化，使生效，批准，证实

sanitarian [ˌsænɪˈteərɪən] *n*. 卫生专家

cross contamination 交叉污染

lubricant [ˈluːbrɪkənt] *n*. 润滑剂，润滑油；*adj*. 润滑的

pesticide [ˈpestɪsaɪd] *n*. 杀虫剂，农药

end-product *n*. 终产品，终产物

agronomy [əˈgrɒnəmɪ] *n*. 农艺学，农学

Hazard Analysis (HA) 危害分析

Critical Control Points (CCP) 关键控制点

microcidal 英 [maikrəubrˈsaidl] 美 [maikroʊbiˈsaidl] *adj*. 杀微生物的

amend [əˈmend] *vt*. & *vi*. 改良，修改，修订

adverse [ˈædvɜːs] *adj*. 不利的，有害的，相反的

chlorine [ˈklɔːriːn] *n*. 氯

sensory [ˈsensəri] *adj*. 感觉的，感官的，传递感觉的

texture [ˈtekstʃə (r)] *n*. 质地，结构

violate [ˈvaɪəleɪt] *vt*. 违反，妨碍，侵犯

audit [ˈɔːdɪt] *n*. 审计，查账；*v*. 审计，查账

efficacy [ˈefɪkəsi] *n*. 功效，效力，效验，生产率

statutory [ˈstætʃətri] *adj*. 法定的，法令的；依照法令的；可依法处罚的

conformity [kənˈfɔːməti] *n*. 符合，一致；遵从；依照

supply chain 供应链

upstream [ˌʌp'striːm] *adj*. 上游的
downstream [ˌdaʊn'striːm] *adj*. 下游的
feasibility [ˌfiːzə'bɪlətɪ] *n*. 可行性，可能性，现实性
compatibility [kəmˌpætə'bɪləti] *n*. 适合，互换性；通用性，和睦相处
operational PRP(s) 操作性前提方案
harmonize ['hɑːmənaɪz] *v*. 和谐

Exercises

Ⅰ. Answer the following questions according to the article

1. What are the seven subparts of cGMP?
2. What are the eight contents of SSOP?
3. How to develop a HACCP plan?
4. What are the preliminary steps for executing a HACCP plan?

Ⅱ. Choose a term to fill in each of the following blanks

1. cGMP is an abbreviation of ________________________. SSOP is an abbreviation of ________________________. HACCP is an abbreviation of ________________________.

2. ________ and ________ are prerequisite programs for HACCP implementation.

3. A full description of the product should be drawn up, including relevant safety information such as: composition, ________, ________, packaging, ________ and ________.

4. Corrective actions are the procedures that are followed when a ________ in a critical limit occurs.

5. ISO ________ are requirements for any organization in the food chain.

6. ISO 9000 family include ________, ________, ________, ________. ISO 22000：2005 family include ________, ________, ________, ________.

7. During hazard analysis, the organization determines the strategy to be used to ensure hazard control by combining the ________, ________.

8. The aim of this International Standard is to ________ on a global level the requirements for food safety management.

参考译文及课后题参考答案

第一单元　食品原料安全控制

1.1　食品安全危害

食源性危害是指食品中存在的引起不利健康影响的生物的、化学的、物理的因素。

1.1.1　生物性危害

生物性危害包括病原性细菌、真菌、病毒、感染性蛋白质、原生动物和蠕虫寄生虫。微生物病原菌可引发两种食源性疾病：感染和中毒。活的病原菌被人体摄入后在体内繁殖引起疾病会造成感染。人体摄入了病原菌产生的毒素会发生中毒。即使没有摄入活的微生物，中毒也会发生。这通常是发生在食物存放时病原菌生长并产生毒素，而随后的加工处理只杀灭了病原菌而并不能破坏毒素的情形下。生物性危害的表现通常包括胃肠道不适、腹泻、呕吐，有时是死亡。美国公共卫生部把水分多的、高蛋白的低酸食品列为潜在危害的食品。高蛋白食品完全或部分由奶或奶制品、全蛋、肉、家禽肉、鱼、贝类、食用甲壳类（虾、龙虾、蟹）组成。烤或煮的土豆、豆腐和其他大豆蛋白食品、加热的植物性食品、生的芽（像苜蓿芽、豆芽）也会产生危害。这些食品可促进感染性或致病性微生物的快速生长。

1.1.2　化学性危害

食品中可能存在着非常广泛的化学危害，它们可能是食品中天然存在的，也可能是在生产中故意或无意添加的。当被摄入后，会引起胃肠道不适、器官损伤和可能致死的免疫学反应。长期摄入含有有毒化学物质的食品会导致包括癌症在内的慢性影响。下面列出了一些化学性危害。

天然存在的化学性危害：过敏原、毒枝菌素；

农业化学物质：杀虫剂、肥料、抗生素；

重金属：铅、镉、汞、砷、铀；

食品加工环境中使用的化学物质：润滑剂、洗涤剂、消毒剂、制冷剂、杀虫剂；

存在于食品包装材料中的化学物质：增塑剂和其他在塑料生产中添加的化学物质。

1.1.3　物理性危害

食源性物理危害通常称为“外来物质”或“外来体”。一般污染源是环境、食品本身、食品加工场所和个人物品。它们在食品中的存在会引起窒息、口腔或体内的划伤，但是很少致死。

泥土和石子是采收过程中典型的环境污染物。一些外来污染物来源于食品本身，像果核、果梗，鱼骨、肉骨或骨头碎片。大量的外来污染物产生于食品加工场所。维修工人可能会把钉子、切断的金属丝、破碎的刀片掉进食品中。生产加工区域中其他装置或器具的玻璃碎片、硬质塑料和木头碎片会混入到食品中。维修工人、生产线上的工人以及从事食品服务业的操作人员使用和穿着的私人物品经常会掉进食品中，这些物品可能包括戒指、铅笔、纸、耳环、鼻环、扣子、温度计、头发和手套。

1.2 生物性危害的控制

一些教育工作者讲到"三个 Ks"是食品生产中生物性危害控制的全面方案，"三个 Ks"包括杀灭微生物、防止微生物生长和防止微生物污染。更科学地阐述这三个方法是：

1.2.1 杀灭微生物

很多悠久的和新型的方法可用来杀灭微生物。这些方法包括像巴氏杀菌、超高温灭菌这样的热处理，以及像辐照、高静压技术和脉冲电场这样的非热加工。大多数情况下，巴氏杀菌是指在常压下进行烹调或加热，通过杀灭病原菌来保护公众健康以及通过杀灭腐败菌来延长食品货架期。一些要灭菌的食品装在金属罐、玻璃或塑料的零售容器中，密封后在121℃或更高的温度下进行加压杀菌处理。食品也可采用超高温瞬时杀菌，例如在140～150℃下杀菌几秒钟。

1.2.2 防止微生物生长

防止微生物在食品中生长的主要控制措施包括冷藏、冷冻、气调。温度低于微生物最低生长温度，微生物不能再生长。普遍接受的最适宜的冷藏温度为等于或小于5℃。冷冻贮藏可进一步延长易腐食品的货架期。商业化生产的冷冻食品通常贮藏在－18℃。一些易腐食品在常压下包装到留有一定顶隙的容器中。其他因素允许的情况下，好氧微生物在这种产品中会自由生长。可通过排除顶隙部位的氧气（真空包装）或充入抑制气体（气调包装）来抑制或防止好氧性微生物的生长。

1.2.3 防止污染

许多潜在的生物性危害可通过防止交叉污染来避免。交叉污染是有害物质通过下列途径污染到食品：

触摸过生的食物（如鸡肉）的手，接着触摸像沙拉配料这样的不需烹调的食品；

接触生的食物的切肉板或抹布的表面，没经过清洗和消毒，就接触即食食品；

生的或污染的食物接触或滴液到熟的或即食的食物上；

在防止交叉污染这点上，食品加工场所的清洗消毒程序、人员操作是至关重要的。

1.3 化学性危害的控制

食品加工场所应当执行一个化学品控制计划以防止产品受到过敏原、贴错标签的或掺假的原料，以及清洁剂和保养剂的污染。就整个的食品安全和质量工作来说，员工培训和认知度是使得食品化学性危害风险最小化的必要因素。例如，这包括产品标签上声明的成分的准确性的确认，对高风险成分的购货收据、贮藏及使用的适当必要的控制措施的履行。质量保证和生产人员必须核实原料被正确地加入到混合器或进行预混合操作。清洁剂、消毒剂、杀虫剂以及其他非食品化学品需要放置在封闭的上锁的区域里，在食品生产期间不要使用。

1.4 物理性危害的控制

控制食品中的物理性危害有三种主要的方法：

杜绝——包括杜绝玻璃、木头对产品的污染，实行员工操作培训和虫害控制方案。

例如，大多数的加工场所严格禁止使用玻璃或易碎的塑料装置、器具，以免碎片可能会进入食品中。在所有的生产区域不应使用木板、木制工具手柄及加工设备。员工不应佩戴首饰，应该穿工作服和戴发罩。员工工作服应当没有口袋，这样像装在口袋中的笔就不会掉进产品中。诱虫灯通过紫外线吸引飞虫，通过使用诱虫灯可使生产场所内部的昆虫活动降至最低。

清除——使用磁体、筛具等装置。

食品生产者将许多类型的磁体用在入料，加工设备和包装操作方面。筛具经常用于从干燥的原料中分离外来杂质。

检查——使用金属探测器、X 射线和光学拣选机。

许多技术可用于检查食品中的外来物质。应用最普遍的是金属探测器，它可用于在线检测或检查包装后的产品。具有图像增强能力的 X 射线装置可用于检查食品中密度大的外来物质，比如肉制品中的骨头渣。使用可见或紫外光的光学技术可检测水果、蔬菜以及坚果的表面缺陷，和外来的植物性杂质、石头等。

课后题参考答案

Ⅰ. Answer the following questions according to the article

1. A foodborne hazard is a biological，chemical，or physical agent in，or condition of food with the potential to cause an adverse health effect.

2. Biological hazards include pathogenic bacteria，fungi，viruses，prions，protozoans and helminthic parasites.

3. These hazards may cause gastrointestinal distress，organ damage and immunological reactions that may result in death.

4. Kill them，keep them from growing and keep them out.

5. For example，prohibit the use of glass or brittle plastic instruments，utensils to avoid the possible entry of glass fragments into the food product. Use sifters and sieves to separate foreign materials from dry food materials. Use X-ray devices to detect dense foreign material inside food products，such as bone chips in meat products.

Ⅱ. Choose a term from what we have learnt to fill in each of the following blanks. Change the word form where necessary

1. biological，chemical，physical
2. Moist，high-protein，low-acid
3. pasteurisation，irradiation
4. Allergen
5. detector

第二单元　食品营养

天然食品和加工食品已经从基础必需品演化成为全球市场范围内快速增长的大宗贸易商品。全球食品贸易的变化反映在膳食方式的改变上，例如，富含脂肪尤其是饱和脂肪而含未精制碳水化合物的高能食品，其消费量大大增长。这些都跟低能耗的生活方式这种改变息息相关，例如久坐不动的生活方式，包括机动化的交通方式，家庭设备机械化程度提高，工作中体力劳动的任务逐渐被淘汰，休息日以放松消遣为主。

此外，在过去的这十来年，世界范围内的工业化、城市化、经济发展和市场全球化也对膳食和生活方式带来了快速的变化。这对于普通民众，尤其是对发展中国家和正在转型的国家来的人们说，他们的健康和营养状况受到了显著影响。营养是能够改变慢性疾病的首要决定性因素，有越来越多的科学证据表明，膳食的改变会对健康生活带来决定性的影响，包含正面的和负面的。

2.1 食物与营养

食物指人类食用的含有一种以上营养素的物品，包括糖类、脂肪、蛋白质和水。几乎所有的食物都来源于动植物，也有一些来源于其他物质例如真菌。真菌和细菌都可以用于发酵食品和盐渍食品的制作。当食物来源丰盛的时候，食物可能被浪费或者视为商品。而当食物短缺的时候，食物是性命攸关的大事，食物配给问题也成为极易引起纷争的问题。世界范围内，粮食生产的增长速度是大于世界人口的增长速度的，但是粮食配给不均、储备有限，甚至恶劣的天气都会导致饥荒。

营养指人体摄取和利用食物过程的总和，包括人体在摄取、消化和吸收的所有物资。营养是食物极其重要的一个方面，尤其是对婴幼儿等弱势群体来说。所以我们用精确膳食评价来进行营养评价，它可以衡量膳食与健康之间关系的变化，能够确定各因素（例如生物因素、环境因素、心理因素）对日常膳食吸收的影响，从而做出相应的变化对策，评估随着时间的推移膳食上的变化和进行膳食干预后带来的改变。然而准确膳食评估依然是个挑战，尤其是当研究对象是孩子们的时候。

2.1.1 营养元素

营养素指具有营养功能的物质，包括蛋白质、脂类、糖类、维生素、矿物质和水。除了矿物元素和水分，其他的所有营养元素都被称为有机化合物，因为它们含有碳元素。

水占据人体重量的60%，它在人体中会持续减少，需要及时补充。糖类、油脂和脂肪是极其重要的能量营养素。蛋白质发挥了重要的功能性，例如质构性、持水性、乳化性、凝胶性和泡沫形成等。

为了维持健康，维生素是除了必需氨基酸和脂肪外必须向动物体少量提供的有机化合物。根据水溶性或脂溶性不同，维生素可以分为两大类：脂溶性维生素有维生素A、维生素D、维生素E、维生素K；水溶性维生素主要有维生素C（抗坏血酸）、烟酸、硫胺素、核黄酸、叶酸、泛酸、维生素B_6、维生素B_{12}、生物素。

根据是否为人类营养所需要及其在人体中是否起代谢作用，矿物质可分为必需和非必需两类。常见的必需矿物元素有钠、钾、钙、磷。常见的非必需矿物元素有铁、碘、锰、锌、氟。

2.1.2 营养素与食品加工

食品加工是将原料转化成供人类食用的食物的方法和技术。不同的烹调方法对于谷类食物的维生素和矿物质损失影响差异较大。矿物质主要是溶水流失，而B族维生素在烹调中的损失来自溶水流失、加热损失、氧化损失、碱处理损失等多种途径，损失率最高的是维生素B_1。

豆腐制品加工中，往往带来矿物质含量的提高。大豆本身含钙较多，而豆腐常以钙盐或镁盐为凝固剂，因此豆腐是膳食中钙镁元素的重要来源。大豆中的微量元素基本上都保留在豆制品中。但是，豆腐加工也有一部分B族维生素溶于水而损失。其中部分原因是加热降解，而大部分是凝固时随析出的水分流失。

膳食中的蔬菜以新鲜蔬菜为主要食用形式，部分蔬菜用来腌制、干制、速冻和罐藏，加工过程通常会引起维生素和矿物质含量的明显变化。尤其是热力会导致维生素C、维生素B_1、叶酸等维生素的分解损失和溶水损失。

肉、禽、鱼等食物在加工中，主要损失水溶性维生素，而蛋白质和矿物质的损失不大。脂肪含量可能因处理方式而有较大的变化。

乳制品是一类营养丰富的食品，合理加工对乳类蛋白质的影响不大，但是其中的维生素、矿物质等会发生不同程度的损失。

2.2 营养教育

肥胖症，糖尿病和心脏病等慢性病的增多现象，反映了生物学、个体行为和环境因素的复杂交互作用。因此人们对于营养教育的重要性有了更清晰的认识。营养教育需要解决偏食的问题以及情感因素、人文因素，例如观念、信仰、态度、含义、社会规范和环境因素等带来的问题。

营养教育已经被定义为任意一种教育策略的组合形式，伴随着环境的支持，旨在促进拥有对食物具有主观选择的能力和有利于健康和福祉的与营养相关的一些行为，营养教育可以通过多种场地实现，涉及多项活动，涵盖个人的、社区的和政策性的。

课后题参考答案

Ⅰ. Answer the following questions according to the article

1. The natural ingredients present in food that give us nourishment are called nutrients.

2. Nutrients might be divided into two general categories based on the amount that we need.

3. Food refers to what humans eat for the consumption of more than one nutrient, which is composed of carbohydrates, fats, proteins and water.

4. calcium; phosphorus; iron; iodine; sodium.

5. The mineral elements are divided into two categories based on the quantity of them that we need.

Ⅱ. Judge whether the sentence is right or wrong

1. × 2. √ 3. √ 4. √ 5. ×

第三单元 食品添加剂

几个世纪以来，诸多成分在各种食物中发挥着有益的作用。例如，我们的祖先使用盐来保存肉类和鱼类，添加药草和香料以改善食物的风味，用糖来保存水果，使用醋液腌制黄瓜。如今，消费者需要并享受着美味、营养、安全、方便、物美价廉的食物供给。这些均应归功于食品添加剂及其技术的不断进步。

用于制作食物的食品添加剂有数千种。如中国食品添加剂数量超过 2000 种，而美国食品添加剂有 3000 多种，其中许多是我们每天在家里使用的原料，如糖、小苏打、盐、香草、酵母、香料和色素。

另外，一些消费者在食品上看到较长的不熟悉的名称而关注起了食品添加剂，觉得这些都是很复杂的化合物。事实上，我们吃的每种食物——无论是刚刚采摘的草莓还是自制的饼干都是含有一定的添加剂成分，正是它们决定了食品的风味、颜色、质地和营养价值。所有食品添加剂均由各国政府和各国际组织严密监管，目的就是为了确保食物能够安全食用和准确标注。

3.1 食品添加剂的概念

广义上讲，食品添加剂是指添加到食品中的任何物质。在法律上，是指“有目的地使用的任何物质，能够或者预期能够变为某种成分，进而影响食物的特性，这种作用可以是直接的，也可以是间接的。”从定义上看，包括了用于食品生产、加工、处理、包装、运输或储存的任何物质。当然，法律上所定义的目的是进行强制批准。因此，本定义不包括一般公认安全的物质（无需政府批准），也不包括食品添加剂规定之前经政府批准使用的物质，以及其他上市前法定批准的可以使用的着色剂和杀虫剂。

直接食品添加剂是指为了特定目的而添加到食品中的物质。例如，黄原胶，用于色拉调味品、巧克力牛奶、面包店馅料、布丁和其他食物以增加质构。大多数直接添加剂在食品的成分标签上均有标注。

间接食品添加剂是由于其包装、储存或其他处理而成为食品的一部分的微量成分。例如，少量的包装物质可能在储存期间进入食物。在允许以该种方式使用以前，食品包装制造商必须向政府证明与食品接触的这些物质都是安全的。

3.2 食品添加剂的作用

消费者认为添加剂在食品中发挥着有益作用，这是情理之中的事。如果我们愿意自己种植、收获和研磨食物，并花费较多时间来烹饪和装罐，能够接受食品腐败可能腐败的风险，有些添加剂则无需使用。然而如今大多数消费者都已离不开食品添加剂所带来的各种工艺、美观和便捷。

食物中添加食品添加剂的作用如下：

(1) 维持或提高食品的安全性和新鲜度

防腐剂能减缓由霉菌、空气、细菌、真菌或酵母引起的食品腐败。除保持食物的质量外，它们还有助于控制可能引起食源性疾病的污染，包括危及性命的肉毒梭菌中毒。抗氧化剂这样的食物保藏剂可以防止脂肪和油以及含有它们的食物变酸或产生异味。保藏剂也能防止空气中的新鲜水果，如苹果，切开后变成棕色。

(2) 改善或维持食品的营养价值

维生素和矿物质（及纤维）常添加到许多食物之中，以弥补饮食中的不足或者加工中的损失或者为了提高食物的营养质量。这种强化和添加有助于减少营养不良的发生。任何添加了营养素的产品必须正确标注。

(3) 改善食品的口味、质地和外观

添加香料、天然和人工香精及甜味剂可以增强食物的口味。乳化剂、稳定剂和增稠剂能够保持食物颜色或改善外观，赋予消费者所期望的食品质地及组织均一。发酵剂能够提升烘焙食品烘烤期间的品质。一些添加剂有助于控制食品的酸度和碱度，而其他成分有助于保持食品的味道和吸引力，降低脂肪含量。

3.3 食品添加剂的种类

食品添加剂主要分为六大种类：保藏剂、营养强化剂、着色剂、调味剂、组织化剂和其他类型食品添加剂。各类食品添加剂详见表1。

（1）保藏剂

食物保藏剂主要有三类：防腐剂、抗氧化剂和抗褐变剂。

防腐剂对延长零食和方便食品的货架期发挥了重要作用，近年来随着微生物食品安全问题的增加，其用途更加广泛。

抗氧化剂用于防止食品中的脂类和/或维生素氧化。它们主要是用来防止自动氧化及由此产生的酸败和异味。既包括天然物质如维生素C和维生素E，还包括合成化学品如叔丁基羟基茴香醚（BHA）和二丁基羟基甲苯（BHT）等各类物质。抗氧化剂特别适用于干燥和冷冻食品的长期保存。

抗褐变剂用于防止食品发生酶促褐变和非酶褐变，特别是干果制品或蔬菜。

（2）营养强化剂

近年来，随着消费者对营养越来越感兴趣和关注，营养强化剂的使用量不断增加。营养强化剂主要包括维生素和矿物质。

维生素（一些情况下也可用作防腐剂）通常被添加到谷类和谷类产品中，以弥补食品加工过程中损失的营养或提高食品的整体营养价值。牛奶中添加维生素D、面包中添加B族维生素均与预防主要营养缺乏有关。

铁和碘等矿物质在预防营养缺乏病方面也具有极其重要的价值。和维生素一样，矿物质也主要添加到谷物制品中。

（3）食品着色剂

着色剂是一种染料或色素，当其添加或应用于食品、药物或化妆品或人体时，能够单独或通过与其他物质反应而赋予颜色。政府负责管理所有食品着色剂，以确保含有食品着色剂的食品能够安全食用，食品中只能含有批准的食品着色剂，并且准确标注。

食品着色剂用于食品之中，原因如下：①抵消因暴露在光、空气、极端温度、水分和储存条件下的颜色损失；②弥补颜色的自然褪色；③增强天然的颜色；④提供有色到无色食品和“趣味”食品。如果没有食品着色剂，可乐不会是棕色的，人造黄油不会是黄色的，薄荷冰淇淋不会是绿色的。因此目前几乎所有我们食用的加工食品中都含有食品着色剂。

在美国，政府将食品着色剂分为认证和豁免认证两大类，两者在批准和使用上都必须遵守严格的安全标准。

认证的食品着色剂是人工合成的，并广泛使用，因为它们能够产生强烈的均匀的色泽，成本低，更容易混合出各种色调。目前批准使用的食品着色剂有九种。认证的食品着色剂一般不会给食品带来不愉快的风味。

豁免认证的食品着色剂包括诸如蔬菜、矿物质或动物等各种天然来源的食品着色剂。天然食品着色剂的成本通常比认证着色剂高很多，有可能给食品带来非预期的风味。这类食品着色剂包括胭脂树提取物（黄色）、脱水甜菜（蓝红色至棕色）、焦糖（黄色至褐色）、β-胡萝卜素（黄色至橙色）和葡萄皮提取物（红色，绿色）。

认证的食品着色剂被分类为色素和色淀两类。色素溶于水中，并制成粉末、颗粒、液体或其他特殊用途形式。它们可用于饮料、干混合物、烘焙食品、甜食、乳制品、宠物食品和其他各种食品。

色淀是色素的水不溶性形式。它比色素更稳定，是含有油脂的产品或含水量很少不足以分溶解色素料的产品的理想选择。典型用途包括包衣片、蛋糕和甜甜圈混合物、硬糖和口香糖。

(4) 调味剂

调味剂是食品中使用最多的添加剂，主要包括三种：甜味剂、天然香料和合成香料以及增味剂。

最常用的甜味剂是蔗糖、葡萄糖、果糖和乳糖，其中蔗糖是最受欢迎的。然而，这些物质通常被归类为食物而不是添加剂。甜味剂常用于低能量或无热量的甜味食物中，如糖精、阿斯巴甜。

除了甜味剂外，还有超过1700种天然和合成物质用来调节食物风味。在大多数情况下，这些添加剂是几种化学物质的混合物，用于替代天然香料。一般而言，化学合成的调味剂与天然调味剂具有相同的化学结构，发挥相同的作用。

增味剂可以放大或改变食物的风味，它们本身不具有任何风味。增味剂常用于日常食物或汤中，以增强对其他口味的感知，如味精、5′-肌苷酸二钠、5′-鸟苷酸二钠等。

(5) 组织化剂

尽管调味剂占人工合成食品添加剂的大部分，但食品中增稠剂在生产中使用量最大。这些添加剂是用来增加或改变食品的整体质地和口感。组织化剂主要包括乳化剂和稳定剂。

乳化剂既包括天然乳化剂如卵磷脂，也包括人工合成乳化剂如单甘酯、双甘酯及其衍生物。乳化剂的主要作用是使香精、油脂均匀地分散在整个食品体系之中。

稳定剂包括天然树胶（如卡拉胶）、天然淀粉和改性淀粉。这些添加剂已使用多年，能够提供产品中所需的组织状态，如冰淇淋的质构，而且现在的干制品和湿制品均使用稳定剂。它们也可用来防止水分蒸发及挥发性油脂变质。

水分保持剂、磷酸盐，常被用来改变含有蛋白质和淀粉的食品的质构，它们在稳定各种乳制品和肉类产品的组织状态方面起重要作用，能够与蛋白和/或淀粉发生反应，提高食品的持水力。

面团改良剂，如硬脂酰乳酸盐、L-半胱氨酸盐，在改善面团条件上有诸多优点：①提高面粉和其他成分的品质变化的耐受性；②提高机械混合面团过程中抵抗力；③更好地保留面团中气体，以降低酵母用量，缩短发酵时间，增加烘焙产品的体积；④使面团中孔隙更加细密均匀，更具弹性质地；⑤更易切片。它们在特定的条件下也可作增稠剂使用。

(6) 其他类型添加剂

除上述种类外，还有许多其他类型添加剂用于食品中，发挥特定而有限的作用。包括各种加工助剂如螯合剂、酶制剂、消泡剂、表面处理剂及各种溶剂、润滑剂和推进剂等。

表1列出了常见食品添加剂的分类、作用、应用及添加剂具体名称。部分添加剂可用于多种用途。

表1　常见食品添加剂

分类	添加剂名称	作　用	应　用
保藏剂	抗坏血酸，柠檬酸，苯甲酸钠，丙酸钙，异抗坏血酸钠，亚硝酸钠，山梨酸钙，山梨酸钾，BHA，BHT，EDTA，生育酚(维生素E)	防止细菌、霉菌、真菌或酵母引起的食物腐败(抗菌剂)；减慢或防止颜色、风味或质地的变化和延迟酸败(抗氧化剂)；保持新鲜度	果酱和果冻、饮料、烘焙食品、腌制肉类、油和人造黄油、谷物、调味品、零食、水果和蔬菜

续表

分类	添加剂名称	作　用	应　用
甜味剂	蔗糖，葡萄糖，果糖，山梨醇，甘露醇，玉米糖浆，高果糖玉米糖浆，糖精，阿斯巴甜，三氯蔗糖，乙酰磺胺酸钾	增加甜度，可增加也可不增加能量	饮料、烘焙食品、甜食、桌面糖、替代品、各种加工食品
着色剂	FD&C 蓝 1 号和 2 号，FD&C 绿 3 号，FD&C 红 3 号和 40 号，FD&C 黄 5 号和 6 号，橙红 B，柑橘红 2 号，胭脂树提取物，β-胡萝卜素，葡萄皮提取物，胭脂红提取物或胭脂红，辣椒粉油树脂，焦糖色素，水果和蔬菜汁，藏红花（注意：豁免食品着色剂不需要在标签标注，但可以简单地称为着色或添加颜色）	弥补由于暴露在光、空气、极端温度、湿度和储存条件下的颜色偏移或损失；弥补自然条件下的褪色；增强自然色；为无色至有色食品和趣味食品提供颜色	糖果、点心食品、人造黄油、奶酪、软饮料、果酱/果冻、明胶、布丁和馅饼馅等诸多加工食品
香精和香料	天然香料，人工香精、香料	添加特定风味（天然的和合成的）	布丁和馅饼馅、明胶甜点混合物、蛋糕混合物、沙拉酱、糖果、软饮料、冰淇淋、烧烤酱
风味增强剂	谷氨酸钠，水解大豆蛋白，自溶酵母提取物，鸟苷酸二钠或肌苷酸二钠	增强食品中已经存在的风味（不提供自己独立的风味）	各种加工食品
营养强化剂	盐酸硫胺素，核黄素（维生素 B_2），烟酸，烟酰胺，叶酸，β-胡萝卜素，碘化钾，铁或硫酸亚铁，α-生育酚，抗坏血酸，维生素 D，氨基酸（L-色氨酸，L-赖氨酸，L-亮氨酸，L-甲硫氨酸）	替代在加工过程中损失的维生素和矿物质（富集），在可能缺乏营养的膳食中添加营养素（强化）	面粉、面包、谷物、大米、通心粉、人造黄油、盐、牛奶、果汁饮料、速溶早餐饮料
乳化剂	大豆卵磷脂，甘油单酯和甘油二酯，蛋黄，聚山梨醇酯，脱水山梨糖醇单硬脂酸酯	使食品中各成分的均匀混合，防止分离；保持乳化产品稳定，降低黏度，控制结晶，保持成分分散，有助于食品溶解	沙拉酱、花生酱、巧克力、人造黄油、冷冻甜品
稳定剂、增稠剂、黏合剂、组织化剂	明胶，果胶，瓜尔胶，角叉菜胶，黄原胶，乳清	产生均匀的质地，提升口感	冷冻甜点、乳制品、蛋糕、布丁和明胶混合物、酱料、果酱和果冻、酱汁
pH 调节剂和酸度剂	乳酸，柠檬酸，氢氧化铵，碳酸钠	控制酸度和碱度，防止腐败	饮料、冷冻甜点、巧克力、低酸罐头食品、发酵粉
发酵剂	烧碱，磷酸二氢钙，碳酸钙	促进烘焙食品的起发	面包和其他烘焙食品
抗结剂	硅酸钙，柠檬酸铁铵，二氧化硅	保持粉末食物自由流动，防止吸湿	盐、焙烤粉、糖粉
保湿剂	甘油，山梨醇	保持水分	切碎的椰子、棉花糖、软糖、甜食
酵母营养素	硫酸钙，磷酸铵	促进酵母的生长	面包和其他烘焙食品
面团强化剂和改良剂	硫酸铵，偶氮二甲酰胺，L-半胱氨酸	生产更稳定的面团	面包和其他烘焙食品
凝固剂	氯化钙，乳酸钙	保持脆性和坚固性	加工的水果和蔬菜，豆制品
酶制剂	乳糖酶，木瓜蛋白酶，凝乳酶，凝乳酶	改性蛋白质，多糖和脂肪	奶酪、乳制品、肉类

3.4 食品添加剂的批准使用

当今，食品添加剂的研究、监管和监测比历史上任何时期都更加严格。以美国为例，食品与药物管理局（FDA）对其安全使用负有主要法律责任。为了销售新的食品添加剂（或在使用已经批准的以一种用途使用的添加剂用于另一种未经批准的用途的之前），制造商或其他赞助商必须首先向食品与药物管理局申请批准。这些申请书必须提供证据以证明这种物质以即将使用的方式使用是安全的。鉴于最近的立法的结果，自1999年以来，间接添加剂已经通过售前通知程序获得批准，需要原申请书所用的同一资料。

当评估食品添加剂的安全性及是否应该获得批准时，政府需要考虑如下因素：

① 该物质的组成和性质；

② 一般消费量；

③ 短期和长期的健康影响；

④ 各种安全因素。

通过评估确定适当的使用水平，包括安全空间，这是允许对预期无害的消费水平的不确定性的因素。换句话说，获得批准的使用水平远低于预期会产生任何不利影响的水平。

由于科学的固有局限性，政府从来不会绝对肯定使用任何物质不存在任何风险。因此，政府会根据现有的最有效的科学成果来确定当添加剂提出使用时是否有确切的对消费者无危害的依据。

若一种食品添加剂获批，政府会颁布法规，包括其可以使用该添加剂的食品类型、最大使用量以及如何在食品标签上标注。

如果有新证据表明已经使用的产品可能不安全，或者如果消费水平发生了变化，需要再次评估，政府可以禁止其使用或进行进一步研究，以确定该使用是否仍然被认为是安全的。

众所周知的法规，如良好生产规范（GMP），将食品中使用的食品添加剂的量限制在实现预期效果所需的最大量以内。

3.5 食品添加剂在产品标签上的标注

食品制造商需要在标签上列出食品中的所有成分。在产品标签上，按照优先顺序列出成分，其中首先标出最大量的成分，然后按用量的降序排列。在美国，食品标签上必须列出经食品与药物管理局认证的食品着色剂（例如，FD&C 蓝色1号或缩写名称，蓝色1号）的名称。但是一些成分可以统称为“增味剂”“香料”“人工调味料”。如豁免认证的食品着色剂，可以统称为“人工色素”，不用标出每个着色剂的名称。在合成的或单一的食品着色剂、香味或香料中的致敏成分可以在成分表中简单地命名为致敏成分即可。

3.6 食品添加剂的风险

尽管食品添加剂带来了诸多好处，但几年来人们仍很关切食用这些物质后可能产生的潜在短期和长期风险。添加剂的批评人士关注使用添加剂的间接和直接影响。至于所提到的许多好处，并不能有足够科学的证据证明某一添加剂是否安全。很少或根本没有关于我们每天食用的添加了添加剂的鸡尾酒的健康风险或联合效应的数据。

对添加剂所描述的间接风险与使用它们的一些好处恰恰相反。虽然人们认为通过使用食品添加剂可以提供更多的选择和多样的食品，但毫无疑问，添加剂也导致了低营养物质的食品供应的增加。这些所谓的垃圾食品，其中包括许多零食，在生活里却可以用来代替食物中的营养食品。最近，食品工业试图通过在零食中添加营养强化剂来解决这一批评，使这些食物成为人体维生素和矿物质的来源。这个问题的长期有效性值得怀疑。显然，需要普及知识来确保消费者选择有营养的食物。然而，一些科学家认为，即使没有直接的营养益处，食物也能提供快乐。

比间接风险更令人关注的是食品添加剂潜在的直接毒性效应，因为添加剂极少产生短期的急性副作用。尽管偶发过食品添加剂直接导致毒性的情况，即使在添加剂可接受的水平上使用，是对添加剂敏感的人也会出现直接和严重的过敏性反应，如对亚硫酸盐及其他添加剂的过敏反应。然而，通过合理的标注应该能够使敏感的个体避开潜在的过敏原。

长期食用添加剂造成的毒理学问题并没有很好地记录在案。首要关注的问题是癌症和生殖，虽然没有直接证据表明添加剂的食用与这些问题的发生有关，然而一些动物试验表明了一些添加剂具有潜在风险。尽管大多数有害添加剂已经被禁用，但仍有一些用于食品之中，典型的添加剂如糖精。

大多数现有的和所有新增的食品添加剂都必须进行广泛的毒理学评价后才能批准使用。尽管人们仍在质疑动物研究的有效性，但科学家们一致认为动物试验确实提供了安全决策所需的数据。

3.7 结语

多年以来食品添加剂用于对食品的保存、调味、混合、增稠和着色，并且在减少消费者出现严重的营养不良方面发挥了重要作用。这些添加剂还有助于确保提供美味、营养、安全、方便、色彩丰富、价格合理的食品，满足消费者一直以来的期望。

食品添加剂受到严格的研究、监管和监测。政府要求首先证明添加剂在其添加水平上是安全的，之后才能用于食品之中，以达到预期的使用水平。此外，因为科学认知不断地更新和检测方法的不断改进，所有食品添加剂都需要接受持续的安全审查，消费者应该对其所吃的食物感到安全。

课后题参考答案

Answer the following questions according to the article

1. Why are Food additives added to food?

To maintain or improve safety and freshness; to improve or maintain nutritional value; to improve taste, texture and appearance.

2. What is a Food Additive?

Any substance the intended use of which results or may reasonably be expected to result—directly or indirectly—in its becoming a component or otherwise affecting the characteristics of any food. Direct food additives and indirect food additives are included.

3. How many types of common food additives are used in foods?

Preservatives, sweeteners, color additives, flavors and spices, flavor enhancers, nutrients, emulsifiers, leavening agents, humectants, firming agents and so on.

4. How a new additive is approved for use in foods?

To market a new food or additive (or before using an additive already approved for one use in another manner not yet approved), a manufacturer or other sponsor must first petition the government for its approval. These petitions must provide evidence that the substance is safe for the ways in which it will be used.

第四单元　食品感官评价

4.1 引言

在消费品工业（食品及饮料、化妆品、个人护理产品、纺织品及服装、医药制品等）当中，食品及饮料部门最早对感官评定表现出兴趣和提供支持。20 世纪 40 年代到 50 年代中期，感官评定得到了美国军需食品及容器研究院的大力支持并资助了针对三军展开的食品接受度研究。很显然，对于军队来说，保证有充足的营养（通过分析膳食或精心制作菜单）并不能确保军人对食品的接受程度。众所周知，风味以及具体产品的接受程度是很重要的。人们集中精力，试图鉴别出什么食品会更受欢迎或更不受欢迎，并且对食品接受度的测量这种基础性问题进行研究。

Arthur D. Little 公司引进了风味剖面法，一种减少对技术专家依赖的定性描述型方法。虽然技术专家概念在过去和将来都颇受关注，但是风味剖面法却以约有 6 位专家的小组代替个人求得一致性的结论。这手段在实验心理学家当中引发了一定的争论，他们关注小组结论这个概念以及个人对这种一致性结论的潜在影响。但是，在当时，这种方法成为感官评定的焦点，并在学科中创立了新的学术方向，后者激励人们对感官过程的各个方面进行更深入的研究与开发。

至 20 世纪 50 年代中期，美国加州大学戴维斯分校开设了一系列有关感光评定的课程，成为少数几个可培训感官评定专业人员的学院之一。早期的研究在开发并评定具体的测试方法方面特别完善。Boggs 和 Hansen 等人对差别型测试方法进行了评估。除差别型方法之外，还有其他测量技术也可作为评定产品接受度的方法。等级排序法及快感标度在 20 世纪 50 年代的中后期较为常见。在这个时期，诸多技术及科学团体，如美国食品科技者学会感官评定分会都组织了很多感官评定及风味测量方面的活动。

20 世纪 60 年代中期至 70 年代，国际上对食品与农业、能源危机、食品组成与原料价格、竞争及全球化市场的关注，都直接或间接地为感官评定提供了发展机会。经过一段漫长而艰难的酝酿期，感官评定已经发展为一门具有特色和为公众所接受的专门学科。

4.2 感官评定的定义

感官评定是一门人们用来唤起、测量、分析及诠释食品及原料当中那些可被人们的视觉、嗅觉、触觉及听觉所感觉到的特征反应的科学学科。鉴别一种产品的接受度的关键因素，并使之定量模型化，通常被认为是任何一种感官评定规程的核心。

该定义明确了感官评定包含所有的感觉。这一点极其重要，也最为人们所忽视，因为在一些场合，感官评定会被人们视为仅仅是一种“味觉测试”，看起来似乎排斥了其他感觉的存在。如果有人被要求对一种特殊的产品属性做出响应，而此时未特别指出要排除产品的香味因素，那么，所获得的色泽响应很可能会被香味以一种不易察觉的方式所影响，响应会因此变得更加复杂，测试结果也可能会出现误判。产品的外观是会影响到一个人对该产品的味觉等方面的响应的。不管人们倾向于相信什么，或者被告知什么，他们对产品的各种响应都是不同感官信息交互作用的综合结果，和响应来源的关系其实并不大。为了避免获得不完整的产品信息，设计实验时必须要把这一点考虑进去。类似“布置一场测试但是要告诉测试人员不要理产品的颜色，因为以后会把颜色改正过来”的申请，极有可能为我们带来“灾难性”的后果。

该定义试图解释感官评定源自于许多不同学科，但同时也强调了感觉是行为的基础。这种对不同学科的涉及可能有助于解释在商业及学术环境中论述感官评定的功能所遇到的诸多困难。涉及的学科包括实验学、社会学、生理心理学、统计学、家政学，如果是食品的话，还包括食品科学与技术方面的应用知识。

正如该定义所涵盖的，感官评定涉及对食品及其他各种原料的感官特性的测量及评定。此外，感官评定还涉及感官评定专业人员对各种响应的分析及诠释，即在产品适用市场的范围内为技术和产品开发的内部小圈子和外部市场建立起联系。这种联系是必需的，因为通过它，工艺及开发专家可以预见产品的变化对市场的影响。此外，他们还相信，不存在任何会导致市场失败的感官缺陷。把感官测试与其他商业功能想链接是必需的，就像感官评定专业人员也应该对市场策略有所了解一样。

感官评定的原理起源于生理学及心理学。得自于感官实验的一些信息可帮助我们更好地认识感官的性质，而这种认识随之又会对测试规程和测量人类受刺激后产生的响应产生重大影响。尽管近几年感官评定的信息来源得到了一定的改善，不过取得更大进展的还是有关感官生理学及知觉过程行为学方面的信息。Geldard 曾指出，经典的“五种特殊感觉”分别为视觉、听觉、味觉、嗅觉及触觉。其中最后一种感觉包括了温度、痛、压力等方面的感觉。

4.3 组织一个感官评定项目

构成一个有效的感官评定项目的十二关键要素如下：已获批的研究目标和主体；界定的项目策划和商业规划；专业的评定人员；评定的硬件设施；运行测试方法的能力；合格测试人员的储备；标准化的测试人员筛选流程；标准化的测试人员工作表现监控流程；标准化的测试申请和测试结果汇报流程；在线数据处理能力；正式的操作手册；后续的计划和研究项目。在这里仅对设施和测试方法做简单介绍。

4.3.1 设施

基本上，感官评定设施肯定需要占用一定面积的空间，所占面积的范围一般在 400～2000$ft^2$❶。不过，测试格和人员的数量却不随分配空间的增大而增加。典型的评定设施分为 6 部分：接待测试者区；6 人的测试格区；准备、等候和储存样品区；小组讨论区；数据处理/记录区域；办公台/办公室区。其大致分配面积分别为 50ft^2，100ft^2，300ft^2，350ft^2，75ft^2，125ft^2。这些空间的分配是相对而言的，会随着产品类型和每天测试容量的变化而改变。理论上，6 个测试格足以满足每年 1000 个的测试容量。

测试区域的一般照明可采用荧光灯，测试格内的照明则需用白炽灯。测试格桌面的无影灯照明强度要达到 100～110ft 烛光强度（或等同强度）。白炽灯应安装在拱盒稍前一点的位置而且微微向测试者方向倾斜，这样可以减弱测试者投射在产品上面的阴影。此外，还要给白炽灯泡装上乳白的散射玻璃外罩来消除聚光效果。不建议采用有颜色的光（例如红色、黄色和蓝色）来作为测试格内的照明。

4.3.2 测试方法

除了配备相应的设施，还要为运行感官评定程序选定工具，包括一些用来评定产品的测

❶ 1ft（英尺）=0.3048m。

试方法（例如用于差别型检验和用于接收型检验的方法）。众多测试方法被分为三大类：差别型、描述型和偏好型。对应的测试类型分别为：区别差异，如成对比较、三角检验；描述性的分析，如风味剖析，定量描述检验；接受度-偏好，如9项喜好分析。

4.4 肉类制品的感官评定指南

本项指南摘自由美国肉类科学协会发布的“肉类的熟制、感官评定及基本仪器的嫩度测定科研使用指南（第1.02版）”这个手册不是每个人做每项研究必须遵循的“标准”，而是如题目所示，只是一个“科研使用指南”。科研人员必须决定用最恰当的方法来解决当前的问题。这里所提到的方法和手段，在大部分情况下是可接受的，并且可被推荐为最适用的。

推荐使用的科研方法所得到的产品可能不会是消费者可接受的最佳的品质。但是，指南中所推荐的方法和手段是为了控制那些不必要的变化，是为了利用最相关的方法来十分准确地回答要解决的问题，并且，当可行时，能够对已发表的研究进行有效的比较解析。

“指南”中的信息包括收集和准备用于感官评定或/和嫩度评定的样品，如牛肉、猪肉和羊肉的肉排，烤肉、肉饼；指南可能也适用于那些经过特定调理、腌制或者粉碎的产品。其他的主题包括产品的处理、熟制方法、感官评定方法及数据处理概述。

在开始试验前，熟制和处理步骤、感官评定的方法及测定的参数都应该确定下来。在方法选择时应该考虑的因素如下：

你的假说是什么？

你想回答什么问题（试验目的）？

所得的结果将做何用？

你预期检出的差异有多大？

同一样品的重复性和不同样品间的差异性有多大？

如果基于预试验的结果发现不同处理方法间感官评定的差异是不可检出的，那么缺失的显著差异可以用判别分析或描述性分析方法来验证。如果感官评定的差异预期是可检出的，则消费者感官评定方法是最合适的。确定消费者评定的差异是否可检出是非常重要的，因为一旦可检出，则可确定这一差异如何影响消费者的可接受性。

定量型感官评定可以分为三个大类：①判别；②描述分析；③消费者。其中判别分析可以用专业训练的评定人员或者非训练的评定人员，这个与评定的目的有关。如果评定的目的是为了获得处理间差异的较高程度的确定性，则首先考虑专业训练的评定人员。专业的评定人员都是经过精心挑选的、高水平培训的，并且比普通消费者吹毛求疵。当用消费者进行判别评定时，消费者不是很苛刻就是很可能无法评定出差异。评定方法应该基于研究的目标而选择。数据应该基于用于试验的样品量来解释。描述性方法需要用专业培训的评定人员。评定人员可以是普通训练的（6～10次训练）到高水平训练（6个月或更多的训练）或专家级训练（10年及更多年限的经历）。随着训练次数和训练经历的增多，评定人员能够察觉到样品间某些特性的更细微的差异。所以，训练的次数应该做好记录，最终的结果要以评定人员训练水平为准进行解释。描述性分析用于对样品内的某个特性进行定量分析。基于测定方法的不同，测定的范围和特性可能会不同，但是每个方法都是用来测定样品感官特性的差异。

消费者评定可以是质量型的也可以是数量型的。质量型的消费者评定方法会阐明消费者对一种或多种产品的观点，并且这种方法所设计的问题比较宽松。这些数据会非常有价值，但是不能够给出明确的可用于统计数据分析的答案。消费者数量型评定能够提供一种衡量消费者观点的方法，这种方法会利用具有一定数量刻度的问卷进行，并且可以转换成能够进行统计分析的数值。该方法可以量化消费者的反应并能够检测结果间的差异。当然了，在中心场所测试（CLTs）所用到的评定的小隔间，被编码的盛放样品的器皿及打分的方法不是人们食用食物时的常规场所。

4.4.1 样品的收集和准备

除非宰前处理或宰后处理影响肉块的大小，否则牛排类、猪排类烤制肉块不应该确定大小。这是一大优点，但是对一定大小的肉块来说熟制方法是否一致是开展实验课题最关键的方面。当然，肉块还应该考虑去除骨头、结缔组织、去除皮下脂肪的程度、肉块厚度、重量、形状等。绞碎的牛肉饼，要求薄一点，应该特别注意控制重量和厚度。并应该关注肉饼加工制作参数的变动性，使其标准化。所有的这些参数应该控制并/或在一定的范围内标准化，以此得到相关的资料和准确的数据。

（1）样品的收集

强烈建议科研人员在进行科学试验确定样品量大小时，咨询统计学家或者进行适当的统计检验。样品量的确定需要对一系列的参数的变异性进行考虑，具体包括：嫩度（剪切力或者感官嫩度），风味，多汁性，或者其他感兴趣的特性。很多数据统计教材都有为某些特定的实验进行样品量大小估测的指导。利用准确的样品量比利用边缘化的样品量在财务上和科学性方面都更加合理，并且后者可能因样品量的问题不能够检测出本身存在的差异。以上的这些测试对于后续感官评定次数的估测及相应的财务支出的估测都十分重要。一个原则是，一项研究中所选择的样品应该在产品中或加工中具有代表性。不同肌肉间正确的取样十分重要。同一部位肉中选择牛排或者猪排作为对象进行不同试验处理时，为了减小试验误差应该进行统计学上随机化取样（假设不考虑同一部位肉不同位置间的差异）或者统计学上区组化取样（假设考虑同一部位不同位置间的差异）。如果一个胴体或者一块肉代表一个处理的一个重复，那么牛排在某一个部位肉中的取样位置需要标准化。比如：从腰脊肉的肋段取第一个牛排用作感官评定，第二块用作剪切力测定（其他的样品也要遵循这个取样顺序来测定相应指标）。很多研究利用背最长肌肉作为评定产量、宰前因素、宰后处理对嫩度和其他食用品质的影响研究，是因为这一部位在胴体中具有最高的价值，并且基本上售卖的形式都是用作干热熟制的牛排或猪排。尽管如此，其他部位肉也受到了非常大的关注，主要在特征分析和市场化方面，所以，可以进行不同处理对多个部位肉的影响分析。并且分析时，应该记住在一些部位肉中取样位置的影响是很重要的。此外，为保证正确的作答，在实验设计的问题中应该涉及肌肉的选择。已经有数据显示同一动物中不同肌肉间的嫩度的相关性趋于相对较低到偏高的程度，所以说，一个部位肉的结果可能不能够代表其他部位肉。

（2）牛排、猪排和肉饼的差异

以下是推荐的牛排、猪排及肉饼的厚度：

牛排（干热加工）：2.54cm；

牛排（湿热加工）：1.9～2.54cm；

羊排（干热加工）：2.54cm；

猪排（干热加工）：2.54cm；

牛肉饼：91.5g，0.95～1.10cm；

牛肉饼：113.5g，1.10～1.27cm。

为了保证牛排、猪（羊）排厚度的一致性，在分切段的时候一定要遵循“指南”。当产品必须要冷冻时，可以利用电锯在冷冻、切割后实现非常好的一致性。在这里列出所有品种的用于烧烤类肉块的大小是不现实的。可以参考最新版的“肉类购买者指南”［Meat Buyer's Guide（NAMI 2015）］及“肉类标准采购规范”（USDA，2010），也可以采用零售或餐饮业常用的肉块大小，来满足实验需求。以下为三个品种的肉用作烧烤时最小的大小和厚度：

牛肉：1.5kg，5cm；

猪肉：1.0kg，5cm；

羊肉：0.5kg，5cm。

现在零售肉中有很小一部分的外部脂肪，表明分割肉块在烹制前将外部的脂肪不全部移除也该大部分去除。还推荐肉类进行真空包装或者包装在透氧率及透水率很低的材料中。

4.4.2 样品的准备及将样品呈送给品评小组

(1) 品评样品的准备

样品准备方法的选择及样品量大小的确定要以实验目标和处理间、处理内的变异性进行确定。非常关键的一点是递呈给每位品评者的样品量必须是标准化的，不仅大小一样、形状一致，而且还要保证温度一致。

① 训练的品评小组评定

为了获得处理间或处理内的中度或者某些时候较高程度的差异，样品一般被切成小立方体。每个品评者会评定来自同一样品不同位置的二到三块样品。对牛排、切片及烧烤类肉块来说，建议的大小为熟制品：1.27cm×1.27cm×厚度。牛肉饼［基于肉饼的大小（91.5g或113.5g）］可以切成6块或者8块pie-形状的样品。即使分切厚一点的或者大一点的肉饼，将其切成立方体可能会造成肉块的破损或者是不能获得相同大小的待品样品。所以，推荐将肉饼切成pie-的形状或者楔子形。

② 消费者小组品定

在日常的用餐过程中，消费者能够从视觉上评判肉的多汁性，也能够通过将样品切成食用的肉块大小对嫩度有个基本的感知。因此，当进行消费者品评时，评定的样品最好足够大而使评定人员能够切割，也因而能够提供一种能够更加贴切的代表消费者用餐的体验。样品的大小应该标准化。次级肉块中的样品位置的效应应该随机化。样品位置效应在方差分析模型中一般作为随机效应。基于实验目的和背景，大多数建议选用1.27cm^3的小立方体。一个样品内嫩度的差异可以通过在一块牛排中随机化选择小立方体得以实现。但是，当品评人员评定一口大小的立方体时，这一结果不能够很好地模拟日常的用餐体验。研究人员应该明白，当他们选用1.27cm^3的小立方体时，他们是放弃了能够影响消费者感知的实际的用餐体验这一方式，也因此，对结果的释义和利用都会有所影响。如果选用小立方体，则实验的设计（如提高每一项处理下的消费者的数量）也应该进行相应改变。

(2) 样品递呈

为了保证所有的品评人员获得的样品都在待测指标的最佳和一致的温度下进行，需要执行标准的呈放方案。推荐的最低的评定温度为60℃（ASTM E1871，2010）并且在多数情况下，样品需要保持在这一温度。对于每项研究来说，保温、分切及递呈样品的流程应该在实际评定之前确定。温度也应该一直进行检测，以此保证样品在实际递呈时的温度处于标准温度，不会太高或太低。评定人员接收到待测样品的温度的差异性也需要了解。因为风味和质构会受到温度和保温时间的影响，所以样品递呈的温度应该保持一致，保温时间也应该保持一致。

理想的条件是，所有的样品在分切后立刻递呈，评定人员会获得来自同一肉排的不同位置的小立方体肉块，或者在进行大块肉的评定时，获得来自同一部位肉的所有处理的肉样。

如果样品在递呈前需要提前保温，则需要进行预实验确定保温的方法不会影响样品的颜色、嫩度、多汁性或风味。几个样品保温的方案如下：

- 带盖子的平底锅或玻璃盘子置于预先加热过的沙子容器中或者放在温热的盘子或加热的烤箱中（49℃）。
- 电热盘上的双层锅。
- 用铝箔纸将样品裹住或者放在带盖的耐热玻璃或玻璃烤盘中，然后放在烤箱中。
- 用预热过的制酸奶的玻璃盘，放在酸奶机或类似的器具中。

① 样品的准备顺序

每个样品出现在每个评定位置的概率应该是一样的，以此减小评定顺序（样品放置位置）引起的偏差。此外，每个样品出现在其他样品之前或之后的几率也应该一致，以此消除移行的影响。William's Square 设计能够实现上述两种平衡。对于处理较多的试验来说，上述两种处理的平衡较难实现。评定的顺序应该用随机号码发生器自动生成，同时顺序应该在后期的模型中作为随机因素来分析。

对于训练的描述性评定小组，评定的第一个样品的偏差及评定的天数的不同可能会是一个问题。标准化的预热样品——通常是能够代表评定研究中最代表性的产品——应该在品评开始时递呈给品评小组进行讨论。这一预热样品能够提高评定人员之间的一致性，提高评定人员的注意力和评定人员的自信心，同时也会在评定开始前对评定人员进行校准。

评定小组的带头人可以利用预热样品作为评价评定小组人员的偏离性及是否缺乏积极性的参考，以此提高评定小组人员的置信性，并提前去除能够影响感官判断的环境因素。

② 每次评定的样品数量

每次评定时应给出的样品数量要包含以下几点的考虑：

- 样品的特点；
- 评审人员的经历；
- 感官的和精神的疲劳感；
- 每个样品要测定的特性的数量。

当考虑样品/评定小组数量及评定小组/评定时间对评定人员对评定牛肉饼的影响时，可以考虑阅读 Bohenkamp 和 Berry（1987）的论文．在进行消费者评定、设置每一个评审小组的样品数量时，应当格外仔细。因为消费者之间对食物的喜爱程度会有很大的不同，评定人员是数据分析模型里最大的误差源。所以，最好所有的消费者能够评定所有的产品。为了尽量减少评定的疲劳感和评定兴趣或注意力的损耗，每一个评定小组评定的样品的数量都要根据样品的类型和评审表中问题的数量进行限定。对未加调料的肉制品来说，评定人员在一次 1 小时的评审中可以评定 6～8 个样品。辛辣的或者调味比较重的产品，如用辛辣的烧烤风味卤制的猪里脊，在每次评审中的数量应该限定为 4 个味觉清除剂，如室温下的蒸馏水或者无盐的饼干都可以将味蕾的疲劳感及风味移行降低到最低。此外，脱脂的意大利乳清干酪是非常有效的辛辣食物的味觉清除剂，而温水或者是苏打水是对高脂产品非常有效的味觉清除剂。所以应该对多种味觉清除剂进行预实验，以此保证味觉会彻底得清除，且不会引起味蕾疲劳感的增加或者影响产品的特性。在样品数量远大于一次评定所能评定的数量时，最好将评定试验设置在几天进行，在每一天的评定中让评定人员品评一小类产品。如果多天的评定不能实现，则可采用局部平衡的、非完全随机设计进行处理，但是评定人员的数量需要增加以达到每一个样品所需要的评定次数。可以参照数据统计的相关教材来进行局部平衡的非完全随机设计。这些设计会使每次评定中每个样品或每个处理与其他样品或处理出现的次数一致。

(3) 感官评定小组参与人员知情同意书

1979 年，一篇由健康、教育和福利部编制的题为“保护人类受试者的伦理原则和指南”的贝尔蒙报告，用来保护用作研究的人类受试者，其中包含了感官评定人员 (NIH，1979)。这个报告通过满足以下三个原则来保护用于实验的人类受试者：

• 对人们的尊重——保护所有人的自主权，礼貌且尊重对待他们，并且允许他们有知情权；

• 利益——最小化研究项目风险同时最大化其利益；

• 正义——确定所有的程序都是合理的、非剥削性的、经过深思熟虑的，公平实施。

在美国，联邦资助的研究和在某些州实施的研究，以及联邦资助的机构以人类为受试者时，如感官评定人员，必须遵守伦理规定，包括获得参与人的知情同意书并由机构审查委员会 (IRB) 进行监督。为获得参与人的知情同意书而准备的表格应该详细地将所有关于评定小组的信息传达给受试者。知情同意书应该包括实验目的和实验设计，谁可以参与，谁实施这一研究，参与人将被问及什么，潜在的风险和不适，可能的利益/报酬等。

课后题参考答案

Ⅰ. Answer the following questions according to the article

1. Sensory evaluation is a scientific discipline used to evoke, measure, analyze and interpret reactions to those characteristics of foods and materials as they are perceived by the senses of sight, smell, taste, touch and hearing.

2. The methods are assigned to three broad categories: discriminative, descriptive, and affective, the responding test types are: difference: paired comparison, duo trio, triangle; descriptive analysis: flavor profile, QDA; and acceptance-preference: nine- point hedonic, respectively.

3. • What is your hypothesis?

• What questions are you trying to answer (test objectives)?

• How will the results be used?

• How large of a difference are you trying to detect?

• How much variability is there within and between samples?

4. • Beef steaks (dry heat): 2.54cm

• Beef steaks (moist heat): 1.9 to 2.54cm

• Lamb chops (dry heat): 2.54cm

• Pork chops (dry heat): 2.54cm

• Beef patties: not<0.95cm or>1.10cm for 91.5-g patties

• Beef patties: not<1.10cm or>1.27cm for 113.5-g patties

5. Every sample should be served in each serving order an equal number of times to reduce any bias related to serving position. Furthermore, every sample should be served before and after every other sample an equal number of times in order to nullify any bias related to carryover effects.

Ⅱ. Choose a term from what we have learnt to fill in each of the following blanks. Change the word form where necessary

1. Being able to identify and quantitatively model the key drivers
2. the behavioral basis of perception
3. fluorescent, incandescent lighting
4. "Research Guidelines"
5. 1.27cm×1.27cm×the thickness
6. usually a sample that represents the typical product being evaluated in the study
7. six to eight
8. room temperature distilled water and unsalted crackers
9. who can participate, who will be conducting the research, what participants will be asked to do

第五单元　食品毒理学

5.1 毒物

毒物可以被定义为能够使生物体系发生有害反应，严重破坏功能甚或引起死亡的任何物质。然而，以此作为毒物的工作定义并不适用，原因很简单，实际上任何已知化学物质只要给予的量足够大，都能引起损伤和死亡。在充足的剂量下，任何化学物质都是有毒的。通过分子氧和膳食金属的含量可以深刻了解剂量的重要性。空气中氧的浓度为21%时，氧是生命必需的，但是，当空气氧分压达到100%时，则引起啮齿动物广泛的肺损伤，且往往造成死亡。铁、铜、锌等金属是必需的营养素，人类膳食中缺乏此类矿物质，会引起特殊的疾病，而摄入过多则造成致命的中毒。

因此，所有的毒效应都是由机体接触的化学物剂量以及化学物质固有毒性所决定的，此外，也取决于生物体的感受性。毒性定义为化学物质引起生物体有害作用的能力。在不同的化学物质之间产生毒效应所需的剂量有很大差别。

5.2 剂量与浓度

影响化学物质潜在毒性的最重要因素是剂量和浓度。请记住，任何物质在一定剂量下都可能是有毒的，相反，毒性作用很强的物质在极低剂量下也可能是无毒的甚至是机体必需的。水为生命之源，一般是不会出现任何急性和慢性毒性作用的，但摄入极大量的水也可能会产生显而易见的毒性作用，有少量文献报道摄入过多的水会产生急性和慢性毒作用，甚至致命的毒作用。研究认为，饮用过量的水（每天以加仑计，1美制加仑=3.78L）导致的死亡原因是细胞和组织水肿。氟化钠是一种急性毒性很强的物质，其 LD_{50} 为35mg/kg，但微量的氟化钠对机体健康很重要。每天1～2mg的氟化钠能促进牙齿的健康，每天3～4mg氟化钠能导致氟斑牙，每天稍大剂量的氟可以导致氟中毒，表现为骨密度增加和骨刺。纯品维生素D经口急性毒性作用很强，LD_{50} 为10mg/kg或 40×10^4 IU/kg，但机体每天需要10pg（400IU）维生素D以维持健康，维生素D缺乏可以导致佝偻病，严重缺乏甚至可以引起死亡。

5.3 毒理学的定义

毒理学是研究化学物质对生物体有害效应的一门科学。它是一门多学科的科学，包含许多不同的研究领域。然而不论专长于毒理学的哪个领域，一个训练有素的毒理学工作者，必然要履行部分的或全部的毒理学职能工作。毒理学有两项基本职能：一是查明化学物质所产生的有害效应的性质，二是评价有害效应在特定接触条件下产生的机会。最终，毒理学的学科目的就是为制定相应的控制措施防止化学物质有害效应提供科学依据。

5.4 毒理学的范畴

5.4.1 描述毒理学

描述毒理学工作者直接参与毒性实验，为安全性评价和管理法规的制定提供资料。这些资料可能仅涉及对人的影响，如药物和食品添加剂即属此类，或者还要考虑对鱼类、鸟类和植物的可能影响，以及可能干扰生态平衡的其他因素。

5.4.2 机制毒理学

机制毒理学工作者关心的是了解、识别化学物质对生物体毒性作用机制。机制毒理学的研究成果对于应用毒理学的许多领域是非常重要的。在危险性评定中，如要证实提供实验动物观察到的一种不良结局是和人类有关联的，机制性研究资料是非常有价值的。如要排除实验动物有害发生于人类的可能性，机制毒理学资料同样是有价值的。机制毒理学研究数据还应用于涉及制造安全化学品，以及中毒与其他许多疾病的合理治疗上。同时毒作用机制的阐明还可以促进基础生理学、药理学、细胞生物学以及生物化学学科的发展。

5.4.3 管理毒理学

管理毒理学工作者的工作职责是，依据描述和机制毒理学工作者提供的资料进行评定，一种药物或其他化学品如果按规定的用途上市，是否仅具有可以接受的低风险。管理毒理学工作者还参与大气、车间空气以及饮用水中许多化学物质容许标准的制定，而这往往需要结合描述毒理学和机制毒理学的基础研究资料，以及风险评估的原则与方法进行综合分析。

5.5 毒理学研究方法与类型

评价化学物质对人类的可能危害，最适合的资料当然是与所关注的接触提交相同的条件下获得的人类资料。然而，这类资料极其缺乏。通常可以获得的人类资料是职业或临床场合下的人群资料。一般来说，至少是在起始阶段，这种场合的接触水平都是属于安全范围以内的。而偶然发生的意外中毒或环境排放事故往往超出这个剂量范围。因此，某特定化学物质的安全评价资料可以有四五种不同的毒理学研究信息来源。这包括基本的动物毒理学实验、非传统型的替代实验（主要为体外方法）、流行病学研究、临床接触实验以及事故性急性中毒。

作为研究化学物质对人体健康可能危害的方法，任何一种没提及的毒理学研究都具有自身的优缺点，没有哪一种单一的方法是完全理想的。因此，只有对方法的优缺点进行比较权衡，才能对各种方法作出评价。表 5.1 分类列举了一些毒理学研究方法的类型及其优缺点。

表 5.1 毒理学研究方法的类型及其优缺点

研究类型	优点	缺点
体内方法	①实验条件便于控制 ②能够评价对整体动物的作用及器官系统的反应 ③可以研究多种不同效应 ④能够阐明作用机制	①动物与人类反应是否一致不确定 ②接触剂量与人类实际情况差别大 ③动物与人类结构与功能方面的种属差异使外推困难

续表

研究类型	优点	缺点
体外方法	①简便经济 ②宿主条件便于控制 ③减少动物使用，合乎伦理准则 ④能够使用人体材料	①不能全面反映完整机体的复杂反应过程 ②无法检测慢性作用与迟发作用
流行病学研究	①直接观察对人体的作用 ②与化学物质诱发的效应相关的真实暴露情况 ③可以检测到全人群敏感谱 ④能够探讨化学物质之间的相互作用	①费时、费力、费钱 ②存在许多混杂因子 ③暴露条件测定不准确 ④测定指标粗略，多为死亡率与发病率 ⑤事后性，难以提供健康保护
人类临床研究	①实验条件受到严格限定与控制 ②研究对象为人类 ③可以通过定期调整接触的条件显现某种结构，易于取得 NOAEL	①可能耗费较大 ②最敏感人群(老、弱、幼者)可能不适宜参与 ③主要限于对安全接触水平及轻微效应的检测 ④通常限于短时间的接触
急性中毒事故研究	①真实的接触情况 ②仅需少量个体 ③较任何其他人类研究都节省经费	①可能缺乏准确的接触资料 ②研究结论不能普遍适用于人类其他的接触情况

5.6 剂量、反应及剂量-反应关系

5.6.1 剂量

剂量是指接触剂量，即给予机体或加入实验体系中的化学物质的数量，以单位体重给予量［一次给药为 g/kg，多次逐日给药为 g/(kg · d)］的形式来表示。然而，剂量不一定都和毒效应成比例，因为毒性取决于所吸收的化学物质的量。剂量可以细分为内剂量、送达剂量和生物学有效剂量，并可作进一步的定义。内剂量（吸收剂量）是吸收进入体内并全身分布的毒物的实际量。送达剂量（靶剂量）是到达靶器官产生可测到的效应的毒物量。生物学有效剂量是指产生反应或健康效应所需的那部分内剂量。

为使剂量这一概念的意义更为明确，（报告剂量时）接触的时间和频率也应加以说明。毒物进入体内的途径主要是胃肠道、肺、皮肤和其他非肠道（注射）途径。动物接触化学物质按时间可划分为四种：急性、亚急性、亚慢性和慢性接触。急性接触时间不超过 24h，亚急性为重复接触不超过 1 个月，亚慢性为 1～3 个月，慢性则超过 3 个月。

5.6.2 两种类型反应——量反应和质反应

从实际角度来看，反应有两种不同的类型：量反应和质反应。量反应是指随着剂量增加而出现效应的连续性改变，如对体重、食物消耗量的影响，酶抑制作用以及心率等生理功能指标的变化。量反应可通过个体或简单的生化体系加以测定。例如，在哺乳动物细胞培养中加入 2,3,7,8-四氯二苯-对二噁英，会引起细胞中一种特异的细胞色素 P450 酶浓度的增加。这种增加出现在一个较大的剂量范围内并和剂量相关。而个体所表现的量反应，可以用刺激性物质与皮肤接触后引起的炎症为例说明：低剂量仅引起轻微炎症；而当剂量增大时，刺激转向炎症且炎症的严重程度不断增加。另外则是质反应，又称全或无反应。质反应仅有两种情况，即所观察的效应或是出现或是不出现。像死亡率和肿瘤发生（率）即为质反应的例子。

5.6.3 剂量-反应关系概念

接触条件和毒效应谱，二者结合形成相互关联的关系，习惯称为剂量-反应关系。描述

该概念最简单的说法是：如果改变机体接触某有毒物质的剂量，那么其反应也将随之发生改变。阐明剂量-反应关系对毒物的研究来说是必需的基础工作。不论选择测定何种反应，生物体系的反应程度和毒物的接触量之间的关系，总是表现为一定的形式，因而剂量-反应关系被认为是毒理学最经典、最基本的概念。

5.7 总结

毒理学是一门多学科的科学，是研究化学物质对生物体有害效应的一门科学。毒理学有两项基本职能：一是查明化学物质所产生有害效应的性质，二是评价有害效应在特定接触条件下产生的机会。任何物质剂量足够大时都是有毒的。

课后题参考答案

Ⅰ. Answer the following questions according to the article

1. Toxicology is the study of the adverse effects of chemicals on living organisms.

2. adverse effects；foreign chemicals/xenobiotics；living organisms

3. Descriptive toxicology；mechanistic toxicology；regulatory toxicology

4. *In vivo* study，*in vitro* study，epidemiological studies，clinical (human) exposure studies，and accidental acute poisonings.

5. The LD_{50} is the experimentally derived single dose of a substance that can be expected to cause death in 50 percent of the animals tested.

Ⅱ. Choose a term from what we have learnt to fill in each of the following blanks. Change the word form where necessary

1. the adverse effects of xenobiotics on living organisms
2. toxicant
3. Dose
4. dose-response relationship
5. NOAEL

第六单元 食品包装

6.1 引言

如今，食品包装在食品工业中扮演着一个重要的角色。包装具有多种功能；它们可以容装、保存和保护产品。外包装应告知消费者关于产品的信息，在包装设计时还应能促进产品销售。包装还具有其他辅助功能，即对于分销商和消费者来说要减少损失、损坏和废弃物，以及便于存储、处理和其他商业操作。

包装技术将大量技术和材料结合在一起，具有两个基本目标：保护产品和展示待售物品。因此，由于某些因素包装由功能性向展示性的发展：激励消费者购买产品以及赋予产品一个适合销售的产品形象。

根据联合国粮食与农业组织（FAO）报告，50%的农产品由于缺乏包装而损失。由于恶劣的天气，物理因素、化学因素和微生物因素等引起的变质而使产品遭受损失。工业化和自然资源的消耗加速了包装的发展，因此包装制造商或包装机械制造商必须适应并预测这种趋势，并意识到只有自动化可以提供必要的灵活性以满足工业需要。在未来几年，食品包装的发展将是至关重要的，这主要是由于新的消费模式和需求创造以及世界人口的快速增长，预计到 2025 年世界人口约为 150 亿。

6.2 食品包装的功能

包装的功能众多，包括保护原料或防止加工食品的腐败变质以及由于外部因素造成的污染。包装是控制潜在具有破坏性的光、氧和水的屏障。它便于使用，能提供足够的存储，能传达信息，并提供产品可能遭窃的证据。它通过以下方式达到这些目的：

- 防止颜色、风味、气味、质地和其他食物品质的损坏。
- 防止生物、化学或物理危害的污染。
- 控制氧气和水蒸气的吸收和损失。
- 便于产品使用，例如将一餐的膳食量装在膳食“盒”（例如，玉米饼）的包装。
- 在使用前提供足够的储存空间，如可堆积、可重复密封、可倾倒。
- 通过防篡改标签防止或指示包装产品的篡改。
- 通过包装标签传递关于营养成分、制造商名称、地址、重量、条形码等信息。
- 加工者应知晓有关营销和包装标准，包括全球对某些颜色和图片符号的接受程度等。

包装本身可以促进销售。它们可以是刚性的、柔性的、金属化的等，并且还可以附带诸如销售信息、健康信息、食谱和优惠券等信息。

6.3 食品包装技术

在过去十年中，包装最重要的附加功能是延长食品的货架期。在食品包装中，基于减少产品周围氧气浓度，在世界范围内有许多新的方法。

（1）真空包装

真空包装将易腐食品以手动或自动的方式置于塑料薄膜包装内，然后通过物理或机械方法从包装内部排出空气，特别是氧气，从而使包装材料与被包装产品表面紧密接触。

以这种方式包装，可以显著地延缓食品的化学和/或微生物的变质，具体取决于包装产品、包装材料的阻隔性能以及空气残余量和储存温度等因素。在多数情况下，真空包装大大延长了食品的保质期。

（2）活性包装

活性包装被定义为“将辅助成分有意地添加到包装材料内部或置于包装顶部空间中以增强包装系统性能的包装”。活性包装包含能够清除氧气的添加剂或保鲜剂；能够吸附二氧化碳、水分、乙烯和/或风味/异味等的吸收剂；能够释放乙醇、山梨酸盐、抗氧化剂和/或其他防腐剂的释放剂；和/或控温剂。

（3）智能包装

智能包装不同于活性包装。它是指能够感知和提供信息的包装。智能包装设备能够感知并提供有关包装食品功能和特性的信息，并可以提供包装是否完整、是否被恶意打开以及产品的安全和质量等信息，并被应用于产品保真、防盗和产品可追溯等领域。智能包装设备包括时间温度指示器（TTI），气体感知染料，微生物生长指示器，物理冲击指示器以及防篡改、防伪和防盗等诸多示例。

(4) 可食用涂层

该技术使用可食用的涂层或膜来作为产品的保护层，如水果的打蜡处理。目前，以蛋白质、淀粉、蜡、脂质、抗菌和抗氧化化合物等为主要成分开发的可食用膜和涂层可以保护食品免受微生物破坏以及质量损失。

(5) 气调包装（MAP）

MAP是指在包装中营造一个有别于空气气氛的气体组成。主要使用气体的性质如下：二氧化碳——抗微生物效应。氧气——气调包装的目的是通过利用氮气和/或二氧化碳替代氧气来将包装顶部空间中的氧气浓度降低至1%～2%，甚至降低至0.2%。另外一种情形，在零售生鲜肉的气调包装中，使用高氧浓度，甚至在80%以上，以延长氧化肌红蛋白的持续时间从而保证肉品表面的樱桃红色（一种消费者乐于接受的外观颜色）。氮气——惰性气体。

在非呼吸食物（不需要氧气）的气调包装中，在大多数情况下是使用具有高二氧化碳含量（>20%）和低氧气含量的气体氛围（<0.5%）和低于5℃的推荐储存温度。在具有呼吸作用的食物，如新鲜水果和蔬菜的气调包装中，一旦内部气氛达到期望的水平，产品的呼吸速率应与气体在包装材料中的扩散速率相匹配，以便维持一个平衡的气调包装氛围。

(6) 无菌包装

无菌包装通常意味着在无菌条件下将热处理后的食物转移到“无菌”的、气密性的容器中，以避免二次污染的发生。其原理以对液体产品如（UHT）牛奶、果汁等产品的包装而著称。食品和包装材料两者均要独立灭菌，并在无菌环境条件下完成包装是无菌包装的原则。目前无菌包装是一种更加主流的包装技术。

在无菌包装系统中，包装材料由聚乙烯、纸板和铝箔组成。它通过热（过热蒸汽或干燥热空气）或热和双氧水的组合进行消毒，然后通过包装机以辊式给料的方式形成标准的砖/块形状（即利乐砖）。

6.4 包装材料的选择

在为其产品选择适当的包装时，包装商必须考虑许多变量。例如，罐头制造商做出包装选择时必须考虑成本，产品相容性，货架期，尺寸灵活性，处理系统，生产线填充和封合速度，反应过程，阻透性，抗凹痕和防篡改性以及消费者的方便性和偏好。

使用薄膜作为其包装材料的加工商必须基于阻隔性能来选择薄膜材料，以防止氧气、水蒸气或光对食物产生不利影响。例如，包装材料的使用能防止光诱导反应以此来控制叶绿素的降解，植物的脱色和红肉的变色，牛奶中核黄素的破坏和维生素C的氧化。为了烹饪的需要，一些薄膜要求具有热稳定性，而一些需要在冷藏或冷冻存储的薄膜则要有耐低温性。最常见的食品包装材料包括金属、玻璃、纸和塑料。

(1) 金属

金属如钢和铝应用于罐和托盘。金属罐可形成气密密封，从而抵抗气体和蒸气进入或逸出，并对内容物提供保护。托盘可以是可重复使用或一次性可回收托盘，尺寸是保温餐桌的尺寸或10号罐的尺寸。金属也用于瓶盖和外包装。

(2) 玻璃

玻璃由金属氧化物如二氧化硅（砂）制备。它用于形成瓶或罐（随后进行密封），以此来防止水蒸气或氧气损失。玻璃的厚度必须足以防止来自内部压力，外部冲击或热应力所导致的破裂。太厚的玻璃增加了重量，从而增加了运输成本，并且受到热应力或外部冲击破坏的可能性会变大。

(3) 纸

纸由木浆制备，并且可能含有其他助剂或成分，例如铝层、塑料涂层、树脂或蜡。这些添加剂使其具有破裂强度（抗爆裂强度）、湿强度（泄漏保护）、耐油性和抗撕裂性，还具有保鲜作用，避免包装食品蒸气蒸发和环境污染，并延长其货架期。

改变纸张的厚度可以实现更结实和更具刚性的包装。

• 纸张较薄（一层）并且具有柔性，通常用于袋子和外包装纸。牛皮纸（德语为“牢固的”）是强度最高的纸。其可被漂白用于肉品包装或保持未漂白用于制作食品杂货袋。

• 纸板较厚（但仍为一层），并且刚性更大。耐烘焙纸板通过用 PET 聚酯（聚对苯二甲酸乙二醇酯）涂覆纸板制成，用于常规烤箱或微波炉。

• 多层纸板构成纤维板，被认为是“硬纸板”。

(4) 塑料

塑料具有可收缩、不可收缩、柔性、半刚性和刚性等几种应用，并且其厚度可变化。使塑料成为包装材料良好选择的重要性质包括以下：

• 柔性和可伸缩性；
• 质轻；
• 低温成型性；
• 耐破裂，具有高破裂强度；
• 热封强度高；
• 阻隔性能多样，如对氧气、水分和光的阻隔。

常见的塑料食品包装材料主要有聚乙烯（PE）、聚丙烯（PP）、聚酯（PET）、聚苯乙烯（PS）、聚氯乙烯（PVC）、聚酰胺（PA）、聚偏二氯乙烯（PVDC）和乙烯-乙烯醇共聚物（EVOH）。

PE 是最常见和最便宜的塑料，占总塑料包装的 63%。它是阻水性材料，防止脱水和冻伤。PP 具有比 PE 更高的熔点和更大的拉伸强度，其通常用作经受较高杀菌温度（例如，蒸煮袋或桶）的食品包装的内层。PET 用于“越来越多的食品和饮料”，包括用作分配食物的管。PET 的一些优点是它耐受高温，并且重量比它替代的玻璃轻。PS 是一种多功能、便宜的包装材料，占总塑料包装的 8%。当发泡时，其通用名称是发泡聚苯乙烯（EPS）。这种发泡聚苯乙烯可用于一次性包装和饮用杯，它提供隔热和防护包装。PVC 占总塑料包装的 6%。PVC 阻挡空气和水分，防止冻伤，并且对气体、液体、香料和气味具有低渗透性。PVC 通过控制脱水防止气味的转移和保持食物新鲜，并且能够耐受高温而不熔化。PA 是通过单体缩合而成的聚合物：由二胺和双羧酸或在同一分子中具有两个官能团的氨基酸缩合而成。PA 性质可以根据其分子量和结晶度使得变化较宽；通常，这类聚合物具有良好的气体阻隔性，耐穿刺性和耐热性。PVDC 是偏二氯乙烯（85%～90%）和氯乙烯的共聚物，以商品名“赛纶”实现了商业化。PVDC 最显著的优点是其优异的阻氧和阻湿性。PVDC 主要用于多层复合膜和容器涂层。EVOH 也具有非常好的阻氧性能；此外，它在多层结构中更常见；根据共聚物中乙烯和乙烯醇的摩尔比，阻隔性能差异显著。

6.5 食品包装的发展趋势

食品包装的一个不断发展的趋势是研究和开发具有高阻隔性能的新材料。高阻隔材料可以减少所需的包装材料的总量，因为它们具有高阻隔性而可以使材料更薄或更轻。方便性也是食品包装开发的一个“热门”趋势。在制造、分销、运输、销售、营销、消费和废弃物处理等方面的方便性将是非常重要和有竞争力的。第三个重要趋势是安全，其涉及公共卫生和针对生物恐怖袭击的安全。这一点是特别重要的，因为即食产品、微加工食品和鲜切果蔬的消费增加，必须从食物链中消除食源性疾病和食物的恶意改变。

食品科学与包装技术及工程开发和消费者研究密切相关。消费者倾向于不断地追求新的功能材料。因此，新的食品包装体系将与食品加工技术，生活方式的改变，政治决策过程以及科学证明等诸多方面的发展相关。

课后题参考答案

Ⅰ. Answer the following questions according to the article

1. The functions of packaging include such purposes as protecting raw or processed foods against spoilage and contamination, facilitating ease of use, offering adequate storage, conveying information, and providing evidence of possible product tampering.

2. Packaging materials should be selected according to a food product's storage and distribution needs. They may eliminate damaging levels of oxygen, light, and temperature, as well as preventing water-vapor loss, while at the same time protecting the food from spoilage and contamination.

3. Answer by yourself based on the type you choose.

4. Answer by yourself based on the example you choose.

5. The effects of three gases are the following: CO_2—antimicrobial effect; O_2—for fresh agricultural products, oxygen is necessary to keep them alive to prolong the shelf-life. For processed foods, oxygen should be removed to avoid the food oxidation; N_2—slows chemical reactions as an inert gas.

6. These characteristics are mainly related to flexible and stretchable property, low-temperature formability, resistant to breakage, heat sealability, barrier properties to O_2, moisture, and light and so on.

7. The development of new packaging materials will be critical to improve the protection of food; the convenience of packaging will play an important role in the future packaging design; food safety issues will be a serious challenge for food packaging.

Ⅱ. Choose a term from what we have learnt to fill in each of the following blanks. Change the word form where necessary

1. contain, preserve, protect.
2. bad weather, physical, chemical, microbiological deteriorations.
3. the composition of the atmosphere, O_2 scavengers, CO_2 emitters.
4. oxygen, water vapor, light.
5. cost, product compatibility, shelf-life, flexibility of size, handling systems.
6. PA, PVDC, EVOH.

第七单元 食品工厂卫生

食品工厂卫生被定义为在洁净和卫生环境下进行食品生产、制备和储存而采取的卫生措施。食品生产设施、加工设备的卫生设计等，是确保食品安全卫生的重要因素。

7.1 厂房结构与设计

厂房建筑物的大小、结构与设计必须便于食品生产的维修和卫生操作。厂房及各种设施必须做到：

① 提供足够的场地安装设备、存放物料，以利于进行卫生操作和食品的安全生产。

② 应采取适当的预防措施以减少食品、食品接触面或食品包装材料受到微生物、化学物质、污物或其他外来物污染的潜在危害。可以通过适当的食品安全控制及操作规范或有效设计，包括将可能发生污染的不同生产加工分开（可采用以下任何一种或数种手段：地点、时间、隔墙、气流、封闭的操作系统或其他有效方法），以减少食品受污染的潜在危害。

③ 采取适当的预防措施以保护露天发酵容器中的散装食品。

④ 结构合理。地板、走道、天花板应易于清扫，保持清洁及维护状况良好。

⑤ 凡是在有害的气体可能污染的食品区域都应安装足够的通风或控制设备，以将各种气体和蒸气（包括水蒸气和各种有害的烟气）减少到最低限度。

7.2 卫生操作

① 一般保养：生产加工企业的建筑物、固定装置及其他有形设施必须在卫生的条件下进行维护和保养，防止食品成为条例所指的劣质食品。

② 用于清洗和消毒的物质、有毒化合物的存放必须做到：用于清洗和消毒的清洗剂和消毒剂不得被有害微生物污染，而且必须在使用时绝对安全和有效。有毒的清洁剂、消毒剂及杀虫剂必须易于识别、妥善存放，防止食品、食品接触面或食品包装材料受其污染。

③ 虫害控制：食品生产加工企业的任何区域均不得存在害虫。看门或带路的狗可以养在生产加工企业的某些区域，但它们在这些区域不得构成对食品、食品接触面或食品包装材料的污染。

④ 食品接触面的卫生：所有食品接触面，包括工器具及设备的食品接触面，均必须尽可能经常地进行清洗，以免食品受到污染。食品生产设备的非食品接触面也应当尽量经常进行清洗消毒，以防止食品受到污染。一次性用品（如一次性用具、纸杯、纸巾）均应存放在适当的容器里，并且必须认真处理、分发、使用和弃置，以防止污染食品或食品接触面。

7.3 卫生设施及控制

每个生产加工企业都必须配备足够的卫生设施及用具，它们包括，但不仅限于：

① 供水：供水必须满足预期的生产加工要求，而且必须来源充足。

② 输水设施：输水设施的设计及安装必须得当，并得到良好的维护，使其能将充足的水输送到厂区所需用水的场所。避免对食品、供水、设施或工器具构成污染，或造成不卫生的状况。

③ 污水处理：污水必须通过适当的排污系统排放，或通过其他有效途径排除。

④ 卫生间设施：每个生产加工企业必须为其员工提供足够的、方便进出的卫生间设施。

⑤ 洗手设施：洗手设施安装的位置必须恰当、方便，同时必须提供适当温度的流动水。同时还必须：做好有效洗手和消毒手的准备工作；提供干手用的卫生（纸）巾或合适的烘干装置。

课后题参考答案

Ⅰ. Answer the following questions according to the article.

1. Food sanitation is defined as the "hygienic practices designed to maintain a clean and wholesome environment for food production, preparation and storage".

2. ① Provide sufficient space for such placement of equipment and storage.

② Permit the taking of proper precautions to reduce the potential for contamination of food, food-contact surfaces, or food-packaging materials with microorganisms, chemicals, filth, or other extraneous material.

③ Permit the taking of proper precautions to protect food in outdoor bulk fermentation vessels by any effective means.

④ Be constructed in such a manner that floors, walls, and ceilings may be adequately cleaned and kept clean and kept in good repair.

⑤ Provide adequate ventilation or control equipment to minimize odors and vapors in areas where they may contaminate food.

3. ① General maintenance.

② Substances used in cleaning and sanitizing.

③ Pest control.

④ Sanitation of food-contact surfaces.

4. ① Water supply.

② Plumbing.

③ Sewage disposal.

④ Toilet facilities.

⑤ Hand-washing facilities.

5. Hand-washing facilities shall be adequate and convenient and be furnished with running water at a suitable temperature. Compliance with this requirement may be accomplished by providing: Effective hand-cleaning and sanitizing preparations; Sanitary towel service or suitable drying devices.

Ⅱ. Choose a right term from what we have learnt to fill in each of the following blanks

1. size, construction, design

2. Floors, walls, ceilings

3. general, control

4. appropriate

5. adequate, convenient, running

第八单元　食品质量与安全管理体系

8.1 引言

食品工业是世界经济最重要的构成之一，对经济、环境以及社会的良性发展具有非常重要的意义。根据NACE的定义，食品工业包括食品、饮料的加工生产和食品供应几个子行业。食品、饮料的加工生产，包括：肉、鱼、果蔬、油脂、乳制品、谷物相关的制品和淀粉类制品、饮料和糖的加工生产。食品供应包括加工后食品的批发、零售以及其餐饮业。食品供应链涉及多种延伸性活动：农业原材料的采购、最终可消费产品的加工和分配。食品工业还涉及许多参与者，如农民、投入品供应商、制造者、包装商、运输人员、出口商、批发商以及最终的消费者，由于他们的兴趣、文化态度和维度的不同而且具有不确定性，使食品工业充满活力和挑战。

在过去十年中尽管食品工业监管严格，但仍然经历了许多危机：疯牛病，李斯特菌污染，禽流感以及最近的马肉丑闻。这些事件最终导致消费者对购买产品的来源和保藏变得越来越敏感。相关部门的物流必须具有一定的响应性、准确性和透明度，从而使消费者重拾信心并且保持信心。标签的使用、国际法规的不断变化以及技术的创新都影响着并改变了食品的供应，从而建立了诸如产品可追溯性、冷链控制或卫生和质量等原则。其内容涉及商品来源、贸易的透明度以及从原料到成品的派送，而反之成品原材料又受到物流管理的影响，特别是转型阶段上游或下游冷链的影响。最终达到的目的是严格控制产品加工和储存中的卫生和质量。条形码、电子商务标准、全球数据同步以及射频识别技术（RFID）都是保障欧盟法规和国际标准的工具。它们提供了一系列食品安全的准则和标准，以确保国际食品贸易的公平、卫生，从而为消费者提供健康食品。

8.2 食品供应链的风险及控制

（1）冷链

食品的安全与卫生在很大程度上依赖于冷链的保障，它贯穿生产者、承运人、分销商和消费者的所有储存和运输的整个链条中。在这个链条中，温度不能超过每类产品的规定温度，如冷冻产品（速冻食品）、超鲜产品（如乳制品）、新鲜产品（如水果和蔬菜）、巧克力或干货（如杂货，饮料和酒）。

冷链可以在如下情形断开：

- 当需要保存温度不同的产品被放在一起存储或运输时；
- 或当食物存放过于拥挤，产品内部不能达到保存温度时；
- 或当车辆未先预冷，产品在波动温度下存储，直到车辆冷却；
- 或在装载或卸载时间过长时，会有冷损失；
- 当食品在运输或储存中需要不同温度时，必须选择最低温度。

（2）可追溯性

食品工业的可追溯性是食品链中所有参与者和利益相关者所关注的一个关键问题。它是指在生产、加工和销售的各个阶段追踪食品的能力，食品、食品中饲料、食用动物或掺入及可能掺入食品的物质及饲料的途径。所有涉及的环节，无论是专业人士、生产者、加工者还是分销商都必须能够识别和解决关键问题，维持法规的遵从性，执行自我监控，同时公共供给机构必须建立和执行卫生控制规定，消费者必须通过所购产品清晰可辨的标签了解产品，并知道制作和存储产品的方法。

为了提高产品的可追溯性，食品部门需要执行国家和国际双重标准。其目的是为了更好地控制危害和降低风险水平，这也是追踪食物中毒或掺假问题根源所必须要求的。

质量可追溯性是将产品相关的物理流动信息以及产品本身的附加信息相结合的一种方法。这些信息可以是成分的性质、数量，原材料的来源、产品之间的关系、原材料、成品等。

食品工业供应链中的相关人员必须确定：

• 收到、处理和发运的货物（产品类型、生产者名称和地址）；

• 供应商和交付的产品；

• 客户和交付给他们的产品；

• 货物的相关信息记录，以及能够从市场上识别、召回的关于产品生产、加工或销售的记录。

可追溯性可以为公众健康提供支持，帮助当局确定污染的原因，或保证消费者安心食用，帮助公司通过销售和市场份额来提高市场竞争力。

8.3 食品安全保障体系

现代食品工业的商品在本质上具有国内和国际双重性质。全世界的消费者习惯了能够获得非本地种植或非应季的食品。他们认为产品的新鲜度、质量和安全是至关重要的。而标准的制定和遵守使每个人都受益。标准为产品在质量和安全性两方面的统一提供了框架，它们降低成本并提高生产效率。国际公认的标准为世界各地的食品自由贸易开放了边界和市场。

食品安全标准的定义体现了其重要意义。所谓标准就是将类似产品分类并采用市场参与者通常理解的一致术语来描述它们的参数。标准提高了市场运行效率，食品生产链中标准无处不在。

许多现有文献表明，良好农业规范（GAP）、关键控制点危害分析（HACCP）和国际标准化组织（ISO）标准是全球食品工业都采用的保证食品质量的基准。它们体现了人类在食品安全管理方面食品质量体系的核心三角革命。以前的研究显示，发展中国家在农业、生产和卫生方面缺乏良好的操作。因此，全程食品安全管理有可能促进这些国家的产品出口。

8.3.1 质量和食品安全管理原则与制度

食品质量安全管理的原则是卫生、预防和降低风险、可靠性、一致性、可追溯性、客户和消费者相关性以及透明度和问责制。这些原则通过各种管理体系得以实施，这些体系部分来自食品行业，如 HACCP，还有一些来自其他领域，如六西格玛管理、质量功能部署和全面生产维护。认证方案通常来自各种体系元素的结合和重组，以适应特定类型的工业需要，通常都是特异性和广泛适用性之间的折中。展望未来，我们可能期望持续推动围绕现有系统制定认证计划，扩大现有计划的适用性，并使整个食品供应链涵盖认证计划。

8.3.2 5S 管理体系

5S 是一种简单的车间管理工具，通过干净、高效和安全的方式提高车间生产率，使其可视化管理以确保引入标准化工作。5S 也是一个组织车间和操作的系统的方法，还是一个整体理念和工作方式。它分为 5 个阶段，以字母“S”开头的不同的日本词命名（整理、整顿、清扫、清洁、修身），因此得名 5S。

这五个不同的阶段（与英语描述）是：

整理：整理、清理、分类；

整顿：整顿、简化、按顺序设置、配置；

清洁：清扫、擦拭、擦洗、清洁和检查；

清扫：标准化、稳定、一致；

修身：保持、自律、习俗和实践。

为了完整性，一些公司添加了第 6 个 S（6S），即安全。有些人看来，安全应该是 5S 步骤的一个组成部分，而不是一个单独的阶段。

8.3.3 良好农业规范（GAP）

良好农业规范（GAP）是一个著名的公共安全标准。GAP体系是一套旨在保障生产和储存的最低标准的农业实践指南。GAP体系重点包括：有害生物管理（农药的最佳使用量）、动物养殖场的粪便处理、水质保持、工人和现场卫生、收获后处理和运输指南等。

GLOBALG. A. P. 源于1997年欧盟良好农业规范，由欧盟零售商生产工作组倡议。与欧洲大陆进行超市合作的英国零售商意识到消费者越来越关心产品安全、环境影响因素以及工人和动物的健康、安全和福利。于是，他们统一了自己的标准和程序，并建立了一个独立的良好农业规范认证体系（G. A. P.）。EUREPGAP标准促使生产者遵守欧洲认可的食品安全标准、可持续生产方法、工人和动物福利、规范水的使用、复合饲料和植物繁殖材料的选择。统一的认证使生产者不再需要每年根据不同的标准进行多次审计，因此为生产者节省了开支。

在接下来的十年中，该认证传遍欧洲大陆。受其全球化的影响，世界各地越来越多的生产商和零售商参与进来，得到了全球的认可。为了反映其全球影响力，实现其国际化G. A. P. 的目标，在2007年EUREPGAP更名为GLOBALG. A. P.。到2008年，GLOBALGAP标准涵盖面已扩大到咖啡，茶叶，牲畜和水产养殖。87个国家的90000多名生产者已获得认证。

GLOBALG. A. P. 认证涵盖：

- 食品安全和可追溯性；
- 环境（包括生物多样性）；
- 工人的健康、安全和福利；
- 动物福利；
- 包括集中作物管理（ICM），集中虫害控制（IPC）；
- 质量管理体系（QMS）以及危害分析和关键控制点（HACCP）。

8.3.4 HACCP体系

HACCP是提高食物链食品安全性的国际公认体系。世界上越来越多的公司采用HACCP来预防、减少或消除潜在的包括交叉污染造成的食品安全危害。

HACCP体系的制定，包括：

- 识别潜在危险；
- 在生产中的特定点实施控制措施；
- 监控和验证控制措施是否符合预期效果。

相比传统的检查程序，HACCP具有如下特点：

- 提供系统的方法确保食品安全；
- 为生产者提供更多的食品安全控制；
- 基于科学，而不是简单的已有经验或主观判断；
- 专注于问题发生之前的预防，相比事后检测效果更好。

一个有效的HACCP体系需要两个关键元素：

（1）良好生产规范（GMP）

GMP是为了控制与工厂人员和食品加工环境相关的危害。GMP的实施创造了一个安全、适宜的食品加工环境。GMP包括程序和监控活动以确保人和场所不存在食品安全隐患。GMP为有效的HACCP计划奠定了基础，而且必须在HACCP计划之前制定和实施。

(2) HACCP 计划

采用 HACCP 计划控制危害的范围：

• 直接相关的产品、成分和流程；

• GMP 无法控制的；

• HACCP 计划预防、消除食品安全危害或将其减少至可接受的水平，包括交叉污染造成的危害。

制定 HACCP 计划步骤如下：

• 描述产品、生产工艺及其危害；

• 分析生产工艺，以确定显著性危害；

• 将控制措施落实到控制食品安全危害的具体步骤；

• 监控控制措施的有效性，如果危害没有得到足够控制，采取行动纠正故障。

HACCP 计划遵循七个核心原则。这些原则由食品法典委员会（CAC）制定。食品法典委员会是由联合国粮食与农业组织（FAO）和联合国世界卫生组织（WHO）创建，其主要工作是制定食品标准、指南和相关文本。

HACCP 计划遵循的七个核心原则如下：

(1) 进行危害分析

这涉及：

• 确定在特定的加工条件下可能影响某产品的危害；

• 收集和评估危害的信息和导致其产生的因素；

• 确定哪些危害对食品安全有重要意义，相关操作必须通过 HACCP 计划解决这些隐患。

(2) 确定关键控制点

关键控制点（CCP）是指在食品加工中食品操作者可控制的一个点、步骤或工序。它是防止、消除食品安全危害或将其降低到可接受水平所必需的点。为了确定工艺中的关键控制点，必须在你的 HACCP 计划中提出能够防止、减少或消除危害的工序。

(3) 建立关键限值

关键限值是将安全产品与不安全产品分开的标准。你必须为每个 CCP 设置关键限值。关键限值必须明确定义并且具有可测量性。

(4) 建立监控体系

监测是对 CCP 的关键限值的测量或观察。所有监测结果必须记录。

(5) 建立纠正措施

如果监测中发现了问题，操作中存在风险或将生产出不安全的食品。HACCP 小组必须制订计划来处理这些风险。而且每一个 CCP 采取的纠正措施必须记录下来：

• 危害重新受到控制；

• 识别和控制所有受影响的产品；

• 防止问题再次发生。

(6) 建立验证程序

验证程序是用来确定 HACCP 体系是否处于正常工作状态。验证包括方法、程序、测试和除了监测外的其他检查。

(7) 建立文件和记录档案

你必须全程记录 HACCP 计划的运行情况，包括上面列出的所有步骤。所有所需的监控和验证记录必须完整和准确。实施 HACCP 计划的主要目的是识别生产过程中可以消除危害（物理、化学或生物）的技术方法。除了关键控制点，生产者必须建立关键控制点的限值、监控方法和纠偏步骤。HACCP 体系要求必须进行验证。验证的目的是确定体系有效的运行，危害得到有效的控制。初级生产几乎是不可能引入 HACCP 体系的，食品加工也是，在这一环节鼓励采用良好生产规范（GMP）进行控制。HACCP 是食品安全的保障体系。它通过建立规则消除食品中出现的危害风险，同时不影响法律的执行。GMP 为每个生产阶段都制定了原则，并且进行了准确的描述。GMP、HACCP 这两个体系互相补充。除了 HACCP 体系，其他食品安全控制体系还有英国零售商协会（BRC）、食品安全体系认证 22000 或国际食品标准（IFS）。

8.3.5 GMP

良好生产规范（GMP）是关于产品质量和安全生产、检测方面的建议和指南。在食品和饮料生产中，GMP 实施的目的是保障消费者安全以及按照原定用途持续高质量生产。厂家多年来所追寻的全面质量管理（TQM）、精益生产和可持续发展都与 GMP 的实施密切相关。越来越多的消费者、零售商和执法部门关注食品生产、销售的条件和操作，这些增加了食品生产企业对相关明确规定的需求，如 GMP 规定的内容。证明 GMP 全面有效实施的最有利的证据可能是在消费者投诉或诉诸法律的情况下，可以减少厂家的责任，并保护企业免受起诉。

“GMP 指南”在 1986 首次由 IFST 推出，该“指南”已获得广泛认可，并成为食品科学家和技术人员不可缺少的参考工具。它规定了确保食品生产过程提供质量一致的产品，无缺陷、无污染，以及使产品尽可能安全。第六版已经完全修正，并且更新了所有最新的标准和指南，特别是关于立法推动的领域，如 HACCP。

该“指南”提供了关于食物和饮料生产、储存和销售的管理手段或技术。为食品教育、培训、食品安全执法人员提供有价值的参考。在学术和工业背景下的食品科学家将重视其精度，政策制定者、监管机构也会发现它在许多重要领域具有不可或缺的指导作用。

8.3.6 ISO

国际标准化组织（ISO）标准是一个国际标准，旨在实现世界贸易统一和防止技术壁垒。ISO 9000 的质量体系的本质是所有活动和处理必须在明确分配责任和权限程序框架内执行。ISO 国际标准为解决当今许多全球性挑战提供了实用工具，从管理全球水资源到改善我们所吃食物的安全性。ISO 国际标准通过确保世界在涉及质量、安全和效率方面问题时使用相同的配方，从而让我们对摄食的产品树立了信心。ISO 有 1000 多个食品专用标准，涵盖了从农业机械到运输、制造和储存的一切。

ISO（国际标准化组织）是世界上最大的自愿性国际标准的开发者，为企业、政府和社会提供了福利。它是一个由 163 个国家的国家标准机构组成的网络。ISO 标准为我们所生活的世界作出了积极贡献。它们确保了诸如质量、生态、安全性、可靠性、兼容性、互操作性、效率和有效性等重要特征，并且考虑了经济成本。它们促进贸易、传播知识、分享技术进步和良好的管理实践。当今，从农场到餐桌供应链的每个阶段的食品都经常跨越国界。ISO 国际标准为我们所摄食的产品树立了信心。

ISO 已经制定了一系列食品安全管理系统标准，可供食品供应链中的任何组织使用。

ISO 包括：

• ISO 22000:2005——总体要求（截至 2010 年年底，已在 138 个国家发布了 18 项 ISO

22000 认证）

- ISO/TS 22002-1:2009——食品制造的具体要求
- ISO/TS 22002-3:2011——农业的具体要求
- ISO/TS 22003:2007——审核和认证机构指南
- ISO 22004:2005——应用 ISO 22000 的指南
- ISO 22005:2007——饲料和食物链中的可追溯性

ISO 22000 系列主要是食品安全管理各个方面的标准：

ISO 22000:2005　食品安全管理总体指南

ISO 22004:2014　ISO 22000 体系的应用建议

ISO 22005:2007　饲料和食品链中的可追溯性

ISO/TS 22002-1:2009　食品生产的特定前提方案

ISO/TS 22002-2:2013　餐饮的特定方案

ISO/TS 22002-3:2011　农场的特定先决方案

ISO/TS 22002-4:2013　食品包装生产的特定先决方案

ISO/TS 22003:2013　审计和认证机构指导

8.4 结论

科学和技术的快速变化、立法的变化以及当前的社会经济与人口的现实问题都显著地影响着我们对食物的选择。今天，全球化使人类能够获得来自世界各地更多种类的食物。我们可以在任何地方采购食品，因此，有时会受到不同的质量标准和（预）制备方法的约束。这相当于给我们增加了额外的风险，需要我们在整个食物链的各个层面进行仔细的管理。制造商和监管机构已经认识到他们的责任，并且充分意识到如果在食品安全管理体系中不采用适当的食品安全措施，会产生多么脆弱和不可预测的污染。重新获得消费者的信任，在利益相关者之间针对可接受的风险水平和有效应对风险的安全措施达成国际共识，这仍然是 21 世纪的主要挑战。发展中国家需要获得 FAO 和 WHO 更多的技术支持，以促进食品安全体系的发展。

课后题参考答案

Ⅰ. Answer the following questions according to the article

1. Food industry is divided into processing and manufacturing of different foods and food supply. The food supply chain links a variety of activities. The food industry also involves multiple players. All of aforementioned factors make food industry a very dynamic and challenging industry.

2. When products that are not to be kept at the same temperature are stored or transported together; Or when the food is too crowded, thus the cold does not penetrate into products depth; Or when the vehicle is not refrigerated in advance, the products can take in temperature until the truck is cooled; Or when loading or unloading is taking too long, there is a rapid loss of cold; When transporting or storing products that can be stored at different temperatures, the lowest temperature must be selected.

3. It aims to improve the traceability, better control hazards and to reduce risk levels and is necessary for tracing the source of a problem of food poisoning or fraud.

4. Hygiene, prevention and risk reduction, reliability, consistency, traceability, customer and consumer relevance, and transparency and accountability are the driving principles.

5. These five distinct phases are (with English descriptions):

Seiri: sort, clearing, classify;

Seiton: straighten, simplify, set in order, configure;

Seiso: sweep, shine, scrub, clean and check;

Seiketsu: standardize, stabilize, conformity;

Shitsuke: sustain, self discipline, custom and practice.

6. GAP systems underline pest management (optimal use of pesticides), manure handling at animal farms, maintenance of water quality, worker and field sanitation, guidelines for post-harvest handling and transportation, among others.

7. Food safety and traceability; Environment (including biodiversity); Workers' health, safety and welfare; Animal welfare; Includes Integrated Crop Management (ICM), Integrated Pest Control (IPC); Quality Management System (QMS), and Hazard Analysis and Critical Control Points (HACCP).

8. Provides a systematic approach to ensuring food safety;

Gives more control over food safety to the processor;

Is based on science, rather than simply past experience or subjective judgment;

Focuses on preventing problems before they occur. This approach yields far better results than trying to detect failures through end-product testing.

9. Regain control of the hazard;

Identify and control all affected product;

Prevent the problem from happening again.

10. Conduct a Hazard Analysis

Determine the Critical Control Points

Establish Critical Limits

Establish Monitoring Procedures

Establish Corrective Actions

Establish Verification Procedures

Establish Record-Keeping and Documentation Procedures

11. Good Manufacturing Practice (GMP) refers to advice and guidance put in place to outline the aspects of production and testing that can impact the quality and safety of a product.

12. GMP covers all aspects of production from the starting materials, premises and equipmentto the training and personal hygiene of staff. Detailed, written procedures are essential for each process that could affect the quality of the finished product.

13. ISO 22000: 2005—Overall requirements

ISO/TS 22002-1: 2009—Specific prerequisites for food manufacturing

ISO/TS 22002-3: 2011—Specific prerequisites for farming

ISO/TS 22003: 2007—Guidelines for audit and certification bodies

ISO 22004：2005—Guidelines for applying ISO 22000

ISO 22005：2007—Traceability in the feed and food chain.

Ⅱ. Choose a term from what we have learnt to fill in each of the following blanks. Change the word form where necessary

1. food and drink processing and manufacturing，food supply

2. meat，fish，fruit and vegetables，oils and fats，dairy，cereal related and starch products，beverages and sugar

3. wholesale and retail distribution of processed food，the catering sector

4. fair practices in the international food trade，guarantee hygiene，healthy food products

5. support to public health，authorities determine the causes of contamination，the companies reassure customers，competitiveness

6. Good Agricultural Practices（GAP），Hazard Analysis of Critical Control Points（HACCP），International Organization for Standardizat ion（ISO）

7. clean，efficient，safe，productivity，visual management，the introduction of standardized working

8. assuring minimum standards for production and storage

9. prevent，reduce，eliminate

10. seven，Codex Alimentarius Commission，Food and Agricultural Organization（FAO），World Health Organization（WHO）of the United Nations

11. point，step，procedure

12. quality standards

13. uniformity，technical barriers

14. ISO 9000-based quality system

15. FAO，WHO

第九单元 食品质量安全国际组织——CAC

自古以来，主管部门就制定了一系列食品标准和法规来保护消费者免受危险、掺假和虚假标识食品的困扰。随着食品保藏和运输技术的发展，20世纪的食品贸易规模迅速扩大。然而，不同国家之间食品法规的冲突或不一致性等因素导致了食品贸易障碍，影响了食品流通。很有必要制定协调统一的食品商品标准，促进食品的跨区域流通。

联合国粮食与农业组织（FAO）和世界卫生组织（WHO）先后举行了一系列专家联席会议来协调世界各国的食品商品标准。1962年在日内瓦召开的FAO/WHO食品标准联席会议确立了这两个机构的合作框架。1963年第16届世界卫生大会批准成立国际食品法典委员会（CAC，简称食典委），负责执行FAO/WHO联合食品标准计划。同年，来自30个国家和16个国际组织的代表参加了在罗马召开的第一次CAC会议。目前，CAC共有188个成员，包括187个成员国和1个成员组织（欧盟，EN）；有240个观察员、56个政府间国际组织、168个非政府组织和16个联合国代表。食典委成员资格向FAO/WHO所有成员国和准成员开放。CAC总部设在罗马，每隔一年在罗马或日内瓦举行一次会议。参会代表以国家为基础。如有必要可召开专门或临时会议。

9.1 食品法典委员会的目的

食品法典汇集了国际已采用的全部食品标准、指南和行为准则，涵盖食品安全（残留物、卫生学、添加剂、污染物等）和质量（产品描述、质量等级、标签和证书）等事项，该法典可作为各国制定食品标准和相关法规的依据。CAC始终致力于开发能用于国内食品监管和国际食品贸易、基于科学原理的、国际认可的标准及相关文本，从而实现保护消费者健康、公平的食品贸易之目标。CAC的另一目的是拟定食品标准和出版食品法典，旨在指导推进相关食品标准定义和一致性要求的构建与阐释。同时，CAC也会促使国际兽医组织（OIE）、国际植物保护公约组织（IPPC）等国际间政府或非政府组织承担的食品标准工作相互协调。

此外，CAC及其附属的专业技术机构还会举行中立性质的会议来讨论在其授权范围内的所有与食品安全和贸易相关的话题。来自政府、消费者团体、产业和学术界的代表就食品安全和贸易问题交换意见，并采用相关标准或相关文本。CAC的非成员国家也可以观察员身份参会。

9.2 食品法典的范围

食品法典是包括食品卫生、食品添加剂、农药/兽药残留、污染物、商品（如牛奶、肉类、水果和蔬菜、加工食品）、标签及其说明、分析和取样方法、进出口检验和认证等方面的标准/指南/行为准则。因此，食品法典更像是与食品相关的纵横交错的标准设置。

9.3 食品法典委员会的组织结构和附属机构

CAC由法典委员会、执行委员会、附属机构和委员会秘书处4个重要组织元素组成，每个组织元素都在完成CAC任务中发挥着重要作用。

执行委员会是委员会休会期间食典委的代表执行机构。执行委员会可就总体方向、战略规划和工作计划向食典委提出建议，研究特殊问题，并通过严格审查工作建议和监督标准制定进展来协助管理食典委的标准制订计划。

法典委员会、FAO/WHO区域协调委员会和特设政府间工作组（IT）是CAC下属的三个附属机构。法典委员会又进一步分为综合主题委员会和商品委员会，其任务是拟定标准草案提交给食典委或处理法典程序。综合主题委员会围绕应用于商品或商品组的问题开展工作。他们提出了应用于一般食品、特殊食品或者食品组的概念和原则，并就商品标准的有关规定进行确认或评价。基于专业科学团体的建议，提出有关消费者健康或安全的主要推荐规范。商品委员会则负责具体的食品标准和食品分类标准。商品委员会必要时才召集开会，一旦工作完成商品委员会就进入休会期或被裁撤。协调委员会负责协调食品标准制订活动和区域标准的发展，确保CAC工作对区域利益和发展中国家的关注点作出回应。协调委员会通常由协调国主办（也可能与其他国家共同主持），主要支出费用（口译和笔译）由FAO和WHO联合设立的预算经费来资助。特设政府间工作组（TF）是针对某一特定具体目标而成立的，职权范围非常有限，任务完成之后即被解散。

位于罗马FAO总部的食典委秘书处主要负责组织CAC会议和执行委员会会议，并帮助下设的附属机构与主持国的秘书处保持密切协调。食典委秘书处是一个不少于20人的专业技术团队，负责处理CAC及其附属机构的所有功能性事务（如准备议程、分发工作文件、起草会议报告）以及更新食品法典。食品法典委员会的组织结构见图9.1。

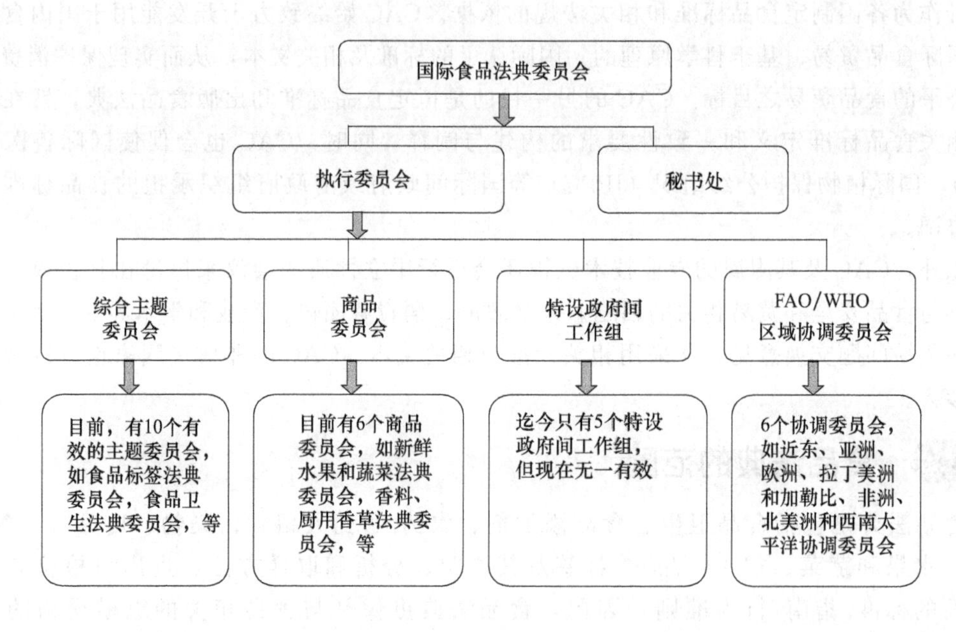

图9.1 食品法典委员会的组织结构

9.4 食品法典委员会如何运行

CAC有一套战略计划来指导其工作，该计划每5年评估一次，主要包含以下5个目标，这5个目标可进一步分解为一系列活动。

- 目标1：促进健全的管理框架；
- 目标2：促进科学原则和风险分析的广泛性和应用一致性；
- 目标3：加强食典委的工作管理能力；
- 目标4：促进国际食典委和相关国际组织间的合作；
- 目标5：促进各成员最大限度的有效参与。

9.5 世界贸易框架下食品法典标准的作用

食品法典作为许多国家食品标准和相关法规的依据，已成为世贸协定中有关食品安全的国际参考文件。这意味着，只要世贸组织成员国将法典标准用于本国的食品安全措施，那么其他成员国将不会挑战这些安全措施作为一种不合理的国际贸易壁垒。如果某个国家想制定严格的食品标准，还需要通过风险性分析程序来使其科学化、合理化。一旦发生争议，法典标准和相关文本也被视为贸易技术壁垒（TBT协议）的国际标准，并用于相关的纠纷解决程序。食品法典标准在TBT协议包含的技术法规和标准条款中承载了相当大的重要意义。TBT协议框架内科学公认的法典标准和相关文本给法典委员会工作带来无穷动力，同时也使其成员产生更多兴趣。已经有越来越多的成员，尤其是发展中国家，参加食典委会议。

9.6 食品法典的现状和未来

现今，消费者享受着来自世界各地的各种食物。有关食品标签、食品添加剂、农药残留、污染物、食品卫生等方面的法典规定为保障食品安全和食品营养成分奠定了基础。因此，无论来自哪里的食品，消费者都对他们所食用的食品安全性和质量更有信心。同时，食品贸易的国际市场不断扩张。越来越多的国家参与到 CAC 标准制定过程中，并在食品生产和加工过程中采用这些标准，从而在促进食品贸易的同时也为世界各国或地区的经济健康发展做出贡献。只要符合这些标准，生产者们就可以确信他们的产品是安全、高品质的，在出口市场是被认可的。因此，越来越多的国家修订他们的食品法律规范，并与食品法典标准保持一致，同时通过加强国内食品监管体系来进一步提高食品质量和安全性。

尽管法典标准在保护健康和促进贸易方面很重要，但由于收入水平较低，许多国家还未能充分参与国际食品安全标准的建立。为改善这种状况，FAO/WHO 在 2003 年发起成立了法典信托基金，旨在协助发展中国家和转型经济国家提高他们有效参与 CAC 活动的能力水平。这也是确保食品法典系统包容性、参与性、公正性的重要步骤。因此，食典委成功的关键在于其成员推动和基于科学、透明、包容、灵活、共识等要素，这种特质使其能够应对成员国提出的任何挑战性问题。今天，食品法典委员仍然面临着诸多挑战，有必要通过努力来鼓励发展中国家提交必要的数据资料并积极参与 CAC 事务，从而使法典标准的制定过程更具包容性。有必要寻找不同社会经济地位和贸易利益/需求的成员之间的共识。有必要增加私有标准的使用，这些私有标准可能与法典标准不同甚至更加严格。在快速变化的环境中，为了保持食品法典的相关性、公平性、包容性和透明性，必须加快脚步制定相关的法规标准。

总之，CAC 强化了其肩负的全球食品产品质量和安全参数确定的重要责任。

课后题参考答案

Ⅰ. Answer the following questions according to the article

1. The purpose of CAC is to making internationally agreed standards and related texts to facilitate domestic regulation and international trade in food commodity. Another purpose is the preparation of food standards and the publication of the Codex Alimentarius, which promotes the coordination of all food standards.

2. The CAC standards includes standards/guidelines/codes of practices for food hygiene, food additives, residues of pesticides and veterinary drugs, contaminants, commodities, labeling and presentation, methods of analysis and sampling, and import and export inspection and certification

3. Fig 9. 1 is expressed the organizational structures and operating mechanism.

4. CAC standards play role in the basis for many national food standards and regulations, international arbitration references for food safety under the WTO's agreement.

5. The CAC must encourage more developing countries to participate fully in a transaction of the CAC.

Ⅱ. Choose a term from what we have learnt to fill in each of the following blanks. Change the word form where necessary

1. Codex Committees, the Joint FAO/WHO Regional Coordinating Committees, Ad hoc Intergovernmental Task Force (TF).

2. food hygiene, food additives, residues of pesticides, veterinary drugs, contaminants, commodities

3. facilitate the domestic regulation and international trade in food fulfill the objectives of consumer health protection and fair practices in food trade

4. internationally adopted food standards, guidelines, and codes of practice

5. international standards, relevant dispute settlements procedures

第十单元 cGMP、SSOP、HACCP、ISO 9000和ISO 22000

10.1 引言

食品安全已成为全球所持续关注的问题，致使许多国家的医疗机构和政府部分寻求方法对生产链进行监控。因此，采用质量控制措施是非常有必要的。这些措施应强调产品和过程的标准化、产品的可追溯性和食品安全保证。食品工业所采用的食品安全系统基础由良好操作规范（GMP）、卫生标准操作程序（SSOP）和危害分析与关键控制点（HACCP）组成。

10.2 前提方案

10.2.1 总则和定义

GMP和SSOP是执行HACCP的前提方案。在工厂中，前提方案主要处理的是“好管家”的问题，而HACCP负责生产过程中具体的危害。工厂必须提供所有的文件，包括所有支持HACCP体系的前提方案的文字记录和结果。重要的是，公司的管理要与系统计划目标相一致，并且与多学科的团队合作来制订和实施计划。检验时，HACCP团队应该对前提方案进行评估，确定这些计划是否继续支持实施HACCP体系时危害分析的决策。

10.2.2 cGMP——现行良好操作规范

现行良好操作规范（cGMP）也称为良好操作规范，是为确保人类或动物消费或使用的产品的安全性而提供的用于生产、检验和质量担保的指南。

现行良好操作规范由七部分组成，其中有两个是预留的（Subpart D & F）（表10.1）。企业可根据实际情况分别执行适合各自需求的良好操作规范。

表 10.1 21 CFR Part 110 总结：在制造、包装或者保存人类食品中的现行良好操作规范

部分	条款	内容	要点
A 部分——总则	Section 110.3	定义	定义： • 酸性食品或酸化食品 • 适当的 • 面糊 • 热烫 • 关键控制点 • 食品 • 食品接触面 • 批 • 微生物 • 害虫 • 厂房 • 质量控制操作 • 返工品 • 安全水分含量 • 消毒 • 必须 • 应该 • 水分活度
	Section 110.5	现行良好操作规范	• 确定劣质食品的标准 • 受特殊的“现行良好操作规范”法规管理的食品也必须符合本法规的要求
	Section 110.10	员工	要求： • 疾病控制 • 清洁卫生 • 教育与培训 • 监督
	Section 110.19	例外情况	• 不属本规范的范围（生的农产品） • 如果有必要将上述例外情况纳入规范时，FDA 会将颁布特别的法规
B 部分——建筑物与设施	Section 110.20	厂房与地面	• 维护地面的方法 • 厂房设计与结构必须便于食品生产的维修和卫生操作
	Section 110.35	卫生操作	要求： • 对器具和设备进行清洗消毒 • 用于清洗和消毒的物质的存放 • 虫害控制 • 食品接触面的卫生 • 已经清洗干净、可移动的设备及工器具的存放和处理
	Section 110.37	卫生设施及控制	要求： • 供水 • 输水设施 • 污水处理 • 卫生间设施 • 洗手设施 • 垃圾及废料的运送
C 部分——设备	Section 110.40	设备及工器具	• 设备和工器具设计、结构和使用的要求
E 部分——生产加工控制	Section 110.80	加工及控制	加工与控制的描述： • 原料与其他辅料 • 加工生产
G 部分——缺陷水平	Section 110.10		• FDA 为天然的或者不可避免的缺陷制定了上限标准 • 虽然食品符合缺陷行动水平，但不能以此作为借口而违反条例 402(a)(4)节的规定 • 不得将含有高于现行缺陷水平的食品与其他食品相混合

注：来源 Federal Register 51，1986。

10.2.3 卫生标准操作规范（SSOP）

卫生标准操作规范是指食品在生产过程中采取卫生措施预防食品被污染或掺假。这些规范必须每天备有证明文件来验证在生产过程中食品是安全的。SSOP应具体到生产食品的每一台设备。根据实际情况，制定自己的SSOP。

食品工厂卫生必须包括便利的环境、加工设备和所有员工。指派一位卫生专家就完全满足食品卫生程序的保证，并负责计划和程序的写作。卫生专家和现场监管人员应该认真地检查生产设备，确保常规的卫生操作已经完成。SSOP应详细说明操作者如何能够满足将要监控的卫生条件和规范。

(1) 卫生控制

每个加工者应在加工前、加工过程中和加工后建立并实施涵盖卫生条件和操作的卫生标准操作程序。SSOP应包括以下内容：

• 与食品或食品接触表面接触的水或用于制冰的水的安全；

• 与食品接触面的状况和清洁度，包括器具、手套及外套；

• 防止不卫生物体对食品、食品包装材料和其他食品接触面，包括器皿、手套和外套以及加工原料对已加工产品的交叉污染；

• 洗手、手消毒和卫生间设施的卫生；

• 保护食品、食品包装材料和与食品接触的表面免受润滑油、燃料、杀虫剂、清洁剂、消毒剂、冷凝物及其他化学、物理和生物污染物的影响；

• 正确标识、储存和使用有毒化学品；

• 控制员工的健康条件，避免对食品、食品包装材料和与食品接触面造成微生物污染；

• 去除食品加工厂内的虫害。

(2) 监控

在加工过程中，加工者应该对卫生条件和操作给予足够次数的监测，以确保它们符合110法规中制定的基本要求。110法规既适用于工厂也适用于被加工的食品。当卫生条件和操作未满足要求时，加工者应及时纠正。

(3) 记录

每个加工者至少应该保存根据规定所作的卫生监控和纠偏行动的SSOP记录。这些记录应该复合第120.12部分中有关记录保存的要求。

10.3 危害分析与关键控制点 (HACCP)

HACCP系统是以科学为基础，通过系统性地确定具体的危害及其控制措施，以保证食品的安全性。HACCP是一个评估危害并建立控制系统的工具，其控制系统是着眼于预防而不是依靠终产品的检验来保证食品的安全。任何一个HACCP系统均能适应设备设计的革新、加工工艺或技术的发展变化。

HACCP适用于从食品的最初生产者到最终消费者的整个食物链。对其实施应建立在对人体健康危险的科学证据指导下进行。除了提高食品的安全性以外，实施HACCP还可以取得其他方面的显著收益。此外，实施HACCP系统还有助于政府部门实施监督，并且由于提高了对食品安全的信任而有助于促进国际贸易。

HACCP的成功应用需要管理部门和从业人员充分参与并承担相应的责任。HACCP的实施还需要采用多学科的知识，必要时应包括农科学、兽医、生产工艺、微生物学、医学、公共卫生、食品技术、环境卫生、化学和工程技术等方面的专业技术人员参与。HACCP与其他质量管理系统，如ISO 9000系列，是兼容一致的。对这些质量管理体系中的食品安全管理方面来说，HACCP是优选的。

制定HACCP计划

在制定HACCP计划时，前5个预备步骤必须在制定针对特殊产品和过程的HACCP计划前完成。

(1) 步骤1　组建HACCP工作组

为建立一个食品加工的有效HACCP计划，应具备与该产品相关的专门知识和专业技能，最好由一个多学科人员组建的工作组来完成。如果现场不能够提供这类专门知识，应从其他途径获得专家的建议。应事先划定该HACCP计划包括食物链中哪些阶段并说明涉及的危害级别（如：是包括所有的危害级别还是仅包括选定的危害级别）。

(2) 步骤2　描述产品

对产品应进行全面的描述。内容包括有关的安全性资料，如成分、物理性质或化学结构(包括水活性，pH等)、微生物杀灭或抑菌处理方法（如：热处理、冷冻、盐渍、烟熏等)、包装、储存期限和储存条件以及销售方式。

(3) 步骤3　确定预期的用途

预期的用途应以食品的最终使用者或消费者所期望的用途而定。特殊情况下，应考虑容易发生健康问题的人群，如集体用餐。

(4) 步骤4　制作流程图

流程图必须由HACCP工作组绘制。流程图应包括整个食品加工操作的所有步骤。当对某项具体操作应用HACCP时，应考虑该操作前后的操作步骤情况。

(5) 步骤5　现场确认流程图

HACCP工作组应在现场对操作的所有阶段和全部加工时段，对照加工过程对流程图进行确认，必要时对流程图作适当修改。

(6) 步骤6　列出每个步骤的所有潜在性危害，进行危害分析，并认定已有的控制措施(原理1)

HACCP工作组应自最初加工开始，对生产、加工、销售直至最终消费的每个步骤，列出所有可能发生的危害，并进行危害分析，以确定在HACCP计划中有哪些危害对食品安全来说是至关重要的，并且必须予以消除或减少到可接受的水平。

进行危害分析时，应尽可能包括下列内容：

- 危害发生的可能性及对健康影响的严重性；
- 危害出现的性质和（或）数量的评估；
- 有关微生物的存活或繁殖情况；
- 毒素、化学物质或物理因素在食品中的产生或残留；
- 以及导致以上情况出现的条件。

HACCP工作组还必须考虑针对所认定的危害已有哪些控制措施。

控制一个具体的危害可能需要采取多个控制措施，而一个控制措施也可能用于控制多个危害。

(7) 步骤7　确定关键控制点（CCP）(原理2)

可能在几个关键控制点上所采取的控制措施都是针对同一个危害的。应用决定树这一逻辑推理方法很容易确定HACCP系统中的关键控制点（CCP)。对决定树的应用应当灵活，在生产、屠宰、加工、储存、销售及其他不同的情况下都可应用。决定树应当被用来指导确认哪些是关键控制点。

如果在某一步骤上对一个确定的危害进行控制对保证食品安全是必要的，然而在该步骤及其他步骤上都没有相应的控制措施，那么，对该步骤或其前后的步骤的生产或加工工艺必须进行修改，以便使其包括相应的控制措施。

(8) 步骤 8　建立关键限值（原理 3）

对每个关键控制点必须制定关键限量并证实其有效性。某些情况下，在一个具体步骤上可能会有多个关键限量。通常关键限量所使用的指标包括温度、时间、湿度、pH、水分活性、有效氯以及感官指标，如外观和质地。

(9) 步骤 9　建立起对每个关键控制点进行监测的系统（原理 4）

监测是有计划地对关键控制点及其关键限量进行测量或观察。通过监测必须能够发现关键控制点是否失控。此外，通过监测还应提供必要的信息，以及时调整生产过程，防止超出关键限量。当监测结果提示某个关键控制点有失去控制的趋势时，就必须对加工过程进行调整。这种调整必须在偏差发生以前进行。对监测数据的分析评价并采取纠正措施必须由具有专门知识并被授权的人员进行。如果监测不是连续进行的，那么监测的数量或频率必须充分确保关键控制点是在控制之下。因为在生产线上没有时间进行费时的分析化验，绝大多数关键控制点的监测程序需要快速完成。由于物理和化学测试简便易行，而且通常能用以指示食品微生物的控制情况，因此，物理和化学测试常常优于对微生物学的检验。另外，所有关键控制点监测的记录和文件必须由监测执行人和另外的审查人员共同签字。

(10) 步骤 10　建立纠偏措施（原理 5）

在 HACCP 系统中，对每一个关键控制点都应当建立相应的纠正措施，以便在出现偏差时实施。

所采取的纠正措施必须能够保证关键控制点重新得到控制。纠正措施还包括对发生偏差时受到偏差影响的食品的处理。出现偏差和受影响食品的处理方法必须记录在 HACCP 文件中保存。

(11) 步骤 11　建立验证程序（原理 6）

建立用于验证的程序。通过验证和审查方法、程序、实验，包括随机抽样及化验分析，可确定 HACCP 是否正确运行。验证的频率应当足以确认 HACCP 系统在有效运行。验证活动的实例包括：

- 审核 HACCP 系统及其记录；
- 审核偏差以及偏差产品的处理；
- 确认关键控制点得到良好的控制。

可能情况下，还应有能够证明 HACCP 计划中所有要素有效运行的证实措施。

(12) 步骤 12　建立文件和记录档案（原理 7）

有效和准确的记录是实施 HACCP 所必需的。HACCP 的实施程序应当用文件规范化，文件和记录必须与食品操作的性质和规模相适应。

文件内容可包括：

- 危害分析；
- 关键控制点的确定；
- 关键限量的确定；

记录可包括：

- 对关键控制点的监测活动；

• 偏差及相应的纠正措施；
• HACCP 系统修改的内容。

10.4 ISO 9000 质量管理体系

ISO 9000 是为了帮助企业为保持有效的质量体系，有效地记录可实施质量体系各要素，而建立的关于质量管理和质量保证的一系列国际标准。

ISO 9000 族涉及质量管理的各个方面并且包含一些 ISO 最知名的标准。这些标准为那些希望确保他们的产品和服务始终如一地满足顾客要求并且质量不断提高的企业和组织，提供了指导和工具。

ISO 9000 族包括：

• ISO 9001：2015　质量管理体系　要求
• ISO 9000：2015　质量管理体系　基础和术语（定义）
• ISO 9004：2009　质量管理体系　业绩改进指南（持续改进）
• ISO 19011：2011　质量和环境管理体系审核指南

10.4.1 ISO 9001: 2015　质量管理体系　要求

ISO 9001:2015 为下列组织规定了质量管理体系要求：

• 需要证实其具有稳定提供满足顾客要求及适用法律法规要求的产品和服务的能力；

• 通过体系的有效应用，包括体系改进的过程，以及保证符合顾客要求和适用的法律法规要求，旨在增强顾客满意度。

10.4.2 ISO 9000: 2015　质量管理体系　基础和术语（定义）

ISO 9000:2015 表述的质量管理的基本概念和原则普遍适用于下列方面：

• 通过实施质量管理体系寻求持续成功的组织；
• 对组织持续提供符合其要求的产品和服务的能力寻求信任的顾客；
• 对在供应链中其产品和服务要求能得到满足寻求信任的组织；
• 通过对质量管理中使用的术语的共同理解，促进相互沟通的组织和相关方；
• 依据 ISO 9001 的要求进行符合性评定的组织；
• 质量管理的培训、评定和咨询的提供者；
• 相关标准的起草者。

本标准给出的术语和定义适用于所有 ISO/TC 176 起草的质量管理和质量管理体系标准。

10.4.3 ISO 9004: 2009　质量管理体系　业绩改进指南（持续改进）

ISO 9004:2009 为组织提供了通过运用质量管理体系方法实现持续成功的指南。它适用于所有组织，无论其大小、类型和活动。

ISO 9004:2009 不是为认证、规范或合同用途而编写。

10.4.4 ISO 19011: 2011　质量和环境管理体系审核指南

ISO 190011：2011 提供了管理体系审核的指南，包括审核原则、审核方案的管理和管理体系审核的实施，也对参与管理体系审核过程的人员的个人能力提供了评价指南，这些人员包括审核方案管理人员、审核员和审核组。

ISO 190011：2011 适用于需要实施管理体系内部审核、外部审核或需要管理审核方案的所有组织。

只要对所需要的特定能力给予特殊考虑，ISO 190011：2011 有可能应用于其他类型的审核。

10.5 ISO 22000: 2005 食品安全管理体系——食品链上组织的要求

ISO 22000：2005 是一个国际标准并定义了食品安全管理体系所覆盖的食品供应链上“从农田到餐桌”所有组织的要求，包括餐饮服务业和包装公司。

标准与一般的关键元素相结合共同确保供应链上食品的安全，包括：相互交流，系统管理，通过 HACCP 前提要求和计划来控制食品安全危害，持续改进和管理系统更新。

ISO 22000：2005 族包括：

- ISO 22000 食品安全管理体系 食品供应链上组织要求
- ISO 22003 食品安全管理体系 为提供食物安全管理体系的审核和鉴定体系要求；
- ISO 22004 食品安全管理体系 ISO 22000：2005 应用指南；
- ISO 22005 饲料和食物链的可追溯性 为体系设计和发展的总的导则；
- ISO 22006 质量管理体系 种植业应用 ISO 9000：2000 指南；
- ISO 2200X 质量管理体系 食品加工和处理组织总的卫生要求。

国际标准制定食品安全体系要求与确定的关键元素结合来确保在食品供应链上的食品安全，直到最终被消费。这些关键元素包括：

- 相互沟通；
- 体系管理；
- 前提方案；
- HACCP 原理。

为了确保食品链每个环节所有相关的食品危害均得到识别和充分控制，整个食品链中各组织的沟通必不可少。因此，组织与其在食品链中的上游和下游组织之间均需要沟通。尤其对于已确定的危害和采取的控制措施，应与顾客和供方进行沟通，这将有助于明确顾客和供方的要求（如这些要求的可行性和必要性以及对终产品的影响）。

为了确保整个食品链中的组织进行有效的相互沟通，向最终消费者提供安全的食品，认清组织在食品链中的作用和所处的位置是必要的。图 10.1 表明了食品链中相关方之间沟通渠道的一个实例。

在已构建的管理体系框架内，建立、运行和更新最有效的食品安全体系，并将其纳入组织的整体管理活动，将为组织和相关方带来最大利益。本标准与 ISO 9001 相协调，以加强两者的兼容性。

ISO 22000 可以独立地应用于其他的管理体系和标准中。它可以结合现有的相关管理体系来要求执行。然而，组织可以利用现有的管理体系来建立一个与国际标准要求相一致的食品安全管理体系。

ISO 22000 国际标准可以与食品法典委员会制定的 HACCP 原理以及应用步骤相结合，根据审核要求，它把 HACCP 计划和前提方案结合在一起。危害分析对于食品安全管理体系的有效性是个关键，因为执行危害分析有助于控制措施与相应的知识有效地结合。标准要求在食品供应链上所有可能发生的危害（包括与加工方式和设备有关的危害）都要被识别并评估，因此它提供了这个方法来决定和证明为什么某些识别的危害需要特殊的组织来控制而其他的不需要。

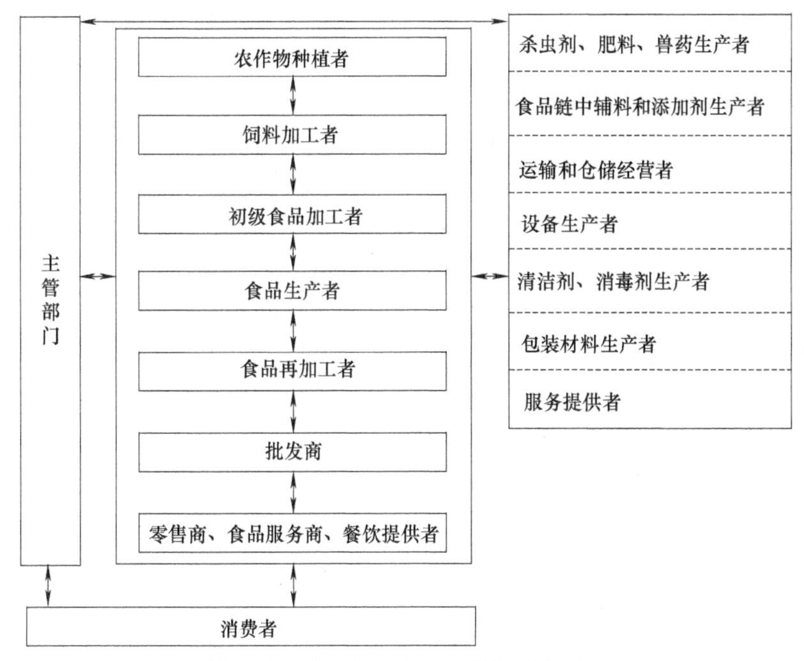

图 10.1　食品链上相关方沟通渠道示例

在危害分析过程中，组织要确定策略并与前提方案、操作性前提方案以及 HACCP 计划相结合，确保控制危害。

为便于这个国际标准的应用，它已发展成一个审核标准。然而，个别组织并没有选择必要的方法来执行这标准的要求。为了帮助个别组织执行这个国际标准，在 ISO/TS 22004 中提供了应用指南。

ISO 22000：2005 国际标准仅关注食品安全。国际标准提供同样的方法来组织和响应食品其他的具体方面（如伦理问题和消费者意识）。

ISO 22000：2005 国际标准允许组织（如规模小的或实力弱的组织）与外部发达的组织联合执行控制措施。

国际标准的目的是将食品供应链上各组织的食品安全管理要求统一在全球水平上。特别是通过组织的应用，寻求一个比法律要求更关注、更一致和更完整的食品安全管理体系。它要求组织通过它的食品安全管理体系来满足与法律法规相关的任何可应用的食品安全。

课后题参考答案

Ⅰ. Answer the following questions according to the article

1. The seven subparts of cGMP are Subpart A general provisions, Subpart B buildings and facilities, Subpart C equipment, Subpart E production and process controls, Subpart G defect action levels, and Subpart D and F are reserved.

2. The eight contents of SSOP are:

• Safety of the water that comes into contact with food or food contact surfaces or that is used in the manufacture of ice;

• Condition and cleanliness of food contact surfaces, including utensils, gloves, and outer garments;

• Prevention of cross contamination from insanitary objects to food, food packaging material, and other food contact surfaces, including utensils, gloves, and outer garments, and from raw product to processed product;

• Maintenance of hand washing, hand sanitizing, and toilet facilities;

• Protection of food, food packaging material, and food contact surfaces from adulteration with lubricants, fuel, pesticides, cleaning compounds, sanitizing agents, condensate, and other chemical, physical, and biological contaminants;

• Proper labeling, storage, and use of toxic compounds;

• Control of employee health conditions that could result in the microbiological contamination of food, food packaging materials, and food contact surfaces; and

• Exclusion of pests from the food plant.

3. In the development of a HACCP plan, there are 12 steps. They are:

• Step 1 Assemble the HACCP team;

• Step 2 Describe the food and its distribution;

• Step 3 Identify intended use;

• Step 4 Construct flow diagram;

• Step 5 On-site confirmation of flow diagram;

• Step 6 List all potential hazards associated with each step, conduct a hazard analysis, and consider any measures to control identified hazards (Principle 1);

• Step 7 Determine critical control points (CCP) (Principle 2);

• Step 8 Establish critical limits (Principle 3);

• Step 9 Establish monitoring procedures (Principle 4);

• Step 10 Establish corrective action (Principle 5);

• Step 11 Establish verification procedures (Principle 6);

• Step 12 Establish record-keeping and documentation procedures (Principle 7).

4. The preliminary steps for executing a HACCP plan are:

• Step 1 Assemble the HACCP team;

• Step 2 Describe the food and its distribution;

• Step 3 Identify intended use;

• Step 4 Construct flow diagram;

• Step 5 On-site confirmation of flow diagram.

Ⅱ. Choose a term to fill in each of the following blanks

1. Current Good Manufacturing Processes, Sanitation Standard Operating Procedure, Hazard Analysis and Critical Control Point

2. CGMP, SSOP

3. physical/chemical structure, microcidal/static treatments, durability and storage conditions, method of distribution

4. deviation

5. 22000

6. ISO 9001: 2015 Quality management systems—Requirements, ISO 9000: 2015 Quality Management Systems—Fundamentals and Vocabulary (definitions), ISO 9004:

2009 Quality management systems—Managing for the sustained success of an organization (continuous improvement), ISO 19011: 2011 Guidelines for auditing management systems, ISO 22000 Food safety management systems—Requirements for organizations throughout the food chain, ISO 22004 Food safety management—Guidance on the application of ISO 22000: 2005, ISO 22003 Food safety management—Requirements for bodies providing audit and certification of food safety management systems, ISO 22005 Traceability in the feed and food chain—General principles and basic requirements for system design and implementation, ISO 22006 Quality management systems—Guidance on the application of ISO 9001: 2000 for crop production data, ISO 2200X Quality management systems—Basic hygiene elements for prerequisite programmes in food producing and handling organization

7. PRP(s), operational PRP(s), the HACCP plan
8. harmonize

Unit 11 Writing of Scientific Thesis

11.1 科技论文的特征、类型及基本结构

11.1.1 科技论文的定义

科技论文是指自然科学领域中用语言文字撰写并公开出版的关于原始科研成果的科学文章。

11.1.2 科技论文的撰写原则

科学性、创新性、逻辑性和通达性是科技论文的灵魂，也是撰写科技论文应遵守的基本原则。

① 科学性指的是论文表述的内容是科学的，另外内容的表述是科学的。

② 创新性指的是论文中揭示出来的事物的现象、属性及其运动规律是前所未见的。

③ 逻辑性指的是论文的前提正确，篇章结构脉络清晰、结构严密、层次分明，推断合理，前后照应，自成体系。

④ 通达性指的是论文中所用词语规范，文字表达通顺流畅，通理达意。

11.1.3 科技论文的类型

严格并科学地对科技论文进行分类是一件很不容易的事情，因为从不同角度考虑就会有不同的分类结果。在高等学校、中国科学院及其他研究单位攻读学位的本科生和研究生为通过学位答辩以获得相应的学位所撰写的论文称为学位论文（Dissertation）。科学技术人员总结自己的科研成果并在科技期刊上公开发表的论文称为科学论文（Scientific Papers）。就科学论文而言，因其写作体裁不同可分为研究论文（Articles）、研究简报（Notes）、研究快报（Letters）和综合评述（Reviews）。

英文科技论文写作是进行国际学术交流必需的技能。一般而言，发表在专业英语期刊上的科技论文在文章结构和文字表达上都有其特定的格式和规定，只有严格遵循国际标准和相应刊物的规定，才能提高所投稿件的录用率。

11.1.4 科技论文的基本结构

在科学技术期刊上公开发表的常见的研究论文（不含快报、简报和综述）的写作格式，按从开头到结尾的顺序包括：文题（Title）、作者姓名及其工作单位（或通信地址）、摘要（Abstract）、关键词（Keywords）、引言（Introduction）、正文（Main Body），其中正文部分包括实验部分或理论方法（Experimental），实验结果或理论结果（Experimental and Theoretical Results），讨论部分（Discussion）、结论（Conclusion），致谢（Acknowledgements），参考文献（References）和附录（Appendix）等。

11.2 文题与关键词的选定

11.2.1 文题的总体要求

论文题目（Title 或 Head-Line）即文章的篇名，简称文题。文题位于论文开篇之首，

是简洁、恰当地反映论文主要内容并对读者具有启迪作用的部分。一个好的文题，应有利于引起读者的兴趣和便于检索、引证，一般基于以下几个原则。

（1）一般性原则

美国化学学会（The American Chemical Society，ACS）在“Style Guide，A Manual for Authors and Editors”（4th printing，1996）一书中，对英文文题提出了如下原则：

① 文题要准确地反映论文的内容。

作为论文的“标签”，题目既不能过于空泛和一般化，也不宜过于烦琐，使人得不出鲜明的印象。

如原有文题（a）Protein in Rice；（b）Applied Package Technology in China。

这都属于用词笼统、空泛，经过修改，以下文题比较恰当：（a）Experiment Research on Protein Loss in Rice；（b）Applied Package Technology in Food Industry of China。

文题应尽量避免使用像“on the”“report on”“regarding”“studies on”“research on”“relationship between…and…”之类的短语。推崇按照文章的主题，直截了当地进行命题。

例如：

On the glass transition of binary blends of polystyrene with different molecular weights 修改为 Glass transition of binary blends of…

在文章的开头，常可省去定冠词 the。例如：

The microstructure of microcrystalline cellulose 修改为 Microstructure of microcrystalline cellulose

The synthesis of a novel alcohol-soluble polyamide resin 修改为 Synthesis of a novel alcohol-soluble polyamide resin

避免使用像 rapid/new 之类非定量、含义不清的词。

② 文题的主题应集中，不求面面俱到。

文题最好不超过 10 ～ 12 个单词，或 100 个英文字符（含空格和标点），如若能用一行文字表达，就尽量不要用 2 行（超过 2 行有可能会削弱读者的印象）。在内容层次很多的情况下，如果难以简短化，最好采用主、副题名相结合的方法，主、副题名之间用冒号（：）隔开。

如 1：Importance of replication in microarray gene expression studies：statistical methods and evidence from repetitive CDNA hybridizations [Proc Natl Acad Sci USA，2000，97 (18)：9834 ～ 9839]，其中的副题名起补充、阐明作用，可起到很好的效果。

如 2：Preparation，thermal decomposition process，non-isothermal kinetics and lifetime equation of binuclear europium benzoate ternary complex 修改为 Preparation，thermal decomposition kinetics of binuclear europium benzoate ternary complex

上例中热分解过程（thermal decomposition process）和寿命方程（lifetime equation）均归于热分解动力学（thermal decomposition kinetics），因而文题可以大大简化。

（2）文题应有新意

例如：Thermal analysis kinetics studies on the dehydration of calcium oxalate monohydrate 可修改为 Application of non-isothermal kinetics of multiple rates to kinetic analysis of dehydration for calcium oxalate monohydrate

水合草酸钙的失水动力学早已有人研究，本文旨在强调应用多重升温速率研究其非等温动力学的研究成果。因此，修改后的文题与正文内容相符，赋有新意。

文题应明确表明研究工作的独到之处，力求简洁有效、重点突出。为表达直接、清楚，以便引起读者的注意，应尽可能地将表达核心内容的主题词放在题名开头。如：The effectiveness of vaccination against in healthy，working adults（N Engl，J Med，1995，333：889-893）中，如果作者用关键词 vaccination 作为题名的开头，读者可能会误认为这是一篇方法性文章：How to vaccinate this population？相反，用 effectiveness 作为题名中第一个主题词，就直接指明了研究问题：Is vaccination in this population effective？题名中应慎重使用缩略语。尤其对于可有多个解释的缩略语，应严加限制，必要时应在括号中注明全称。对那些全称较长，缩写后已得到科技界公认的，才可使用。为方便二次检索，题名中应避免使用化学式、上下角标、特殊符号（数字符号、希腊字母等）、公式、不常用的专业术语和非英语词汇（包括拉丁语）等。

（3）词序

由于题目比句子简短，并且无需主、谓、宾，因此词序也就变得尤为重要。特别是如果词语间的修饰关系使用不当，就会影响读者正确理解题目的真实含意。例如：Isolation of antigens from monkeys using complement-fixation techniques，可使人误解为“猴子使用了补体结合技术”。应改为：Using complement-fixation techniques in isolation of antigens from monkeys，即“用补体结合技术从猴体分离抗体”。

11.2.2 文题的结构特征

（1）文题采用偏正名词短语

常采用的中心词：preparation，synthesis，polymerization，reaction，modification，separation，design，model，method，characterization，identification，analysis，comparison，evaluation，assessment，structure，composition，property，behavior，magnetic properties，irradiation degradation，activity，oxidation，thermostability，compatibility，application，progress，study，等等。中心词会因学科而异。

（2）中心词的限定

在中心词前可加以限定、修饰，通过介词 of、to、in 构成后置定语，或通过 using、under 等引出采用的实验手段、实验条件等。当 study on 之类词前有限定成分时，则应予以保留。例如：

MTDSC and atomic force microscopy studies of morphology and recrystallization in polyesters including oriented films.

但有介词 by 时，应保留 investigation 之类的词，例如：

Investigation of phase behavior of polymer blends by thermal methods

从语法修饰的角度，不应将一长串词作一个词的修饰词。例如：

Thermal decomposition kinetics of the complexes of light rare earth bromides with alanine by DSC method 修改为 DSC study on kinetics of thermal decomposition for the complexes of light rare earth bromides with alanine

此题可在 thermal decomposition 前加介词 of，将其修改为 kinetics 的后置定语。

通常，文题用短语，而不用句子。图、表和分节标题也应避免使用无主语句。可将应用列表、用 DSC 测试热效应、用 TGA-MS 分析分解产物、检查材料性能分别修改为：应用一览表、热效应的 DSC 测试、分解产物的 TGA-MS 分析、材料性能检测。

11.2.3 英文文题基本模式

以“中心词”为基础而展开的基本结构模式是：在中心词前面有限定词，通过介词 of，

to，between 等在中心词后引出后置定语，再通过介词 by，at，in，with，under，over，onto 等补充说明实验方法和条件。

11.2.4 几点具体说明

（1）字数

文题组成字母通常不超过 12 个单词（或 10～15 个词，或 75 个字母以内）。

（2）字母大小写

字母大小写 3 种形成：

① 全大写，例如：

PHASE BEHAVIOR OF POLYMER BLENDS

② 每个实词的首字母大写，但冠词、连词和通常由 4 个以下字母组成的介词（如 the、and、in、at、on、by、via、for、from、over、with、near 等）小写。例如：

Intermacromolecular Complexation Due to Specific Interaction

Calorimetric Study of the Relationship between Molecular Structure and Liquid- crystallinity of Rod-like Mesogens

③ 题名第一个单词的首字母大学，其余小写，常用于目录（CONTENTS）/文献（REFERENCE），例如：

Food organoleptic investigation.

（3）标点

英文题名，主、副题名以冒号连接。例如，Acta Pharmacologica Sinica：Striding Forward Towards World（《中国药理学报》迈步走向世界）。

11.3 摘要

11.3.1 摘要的分类

用英文撰写的科技论文摘要（Abstract Writing）和其他文种撰写的科技论文摘要是一样的，只是文种的区别而已。关于摘要的分类问题，国内部分学者与国外学者虽有分歧，但大同小异。如《科技书刊标准化 18 讲》把摘要分为报道性摘要（Informative abstract）、指示性摘要（Indicative abstract）、报道-指示性摘要（Informative-Indicative abstract）和结构式摘要（Structured abstract），而国外学者一般将摘要分为两种，报道性摘要（Informative abstract）和指示性摘要（Indicative abstract）。

（1）报道性摘要（Informative abstract）

报告性摘要概括了文章的范围和内容，包括研究目的与范围；简述使用材料与方法；简述研究结果；阐明结论与意义。由于报道性摘要的固有特性，特别适合于实验性研究。大多数的学术性杂志论文都是使用报道性摘要，一般科技论文都需要有 100～250 个单词的摘要。

例如：

Low field nuclear magnetic resonance (NMR) technique was used to evaluate the quality and behavior of water in navel orange during the course of storage. Carr-Purcell-Meiboom-Gill (CPMG) sequence was used to acquire the samples'NMR echo signals. The spin-spin relaxation time (T_2) of samples were calculated by a mult-exponential regress model during the deferent period of storage until decay. A trend of T_2 variation was observed in the diagram T_2 vs. storage time. T_2 values arise in the initial stage and then changed little in the following steady period and fall quickly at the end before the samples decay during the period of storage. A T_2 declining quickly processing before decay were observed in nearly all samples. This method would be helpful for evaluating the quality of navel orange and the effect of preserving agent during storage in the future.

实验目的与范围——Low field nuclear magnetic resonance (NMR) technique was used to evaluate the quality and behavior of water in navel orange during the course of storage.

采用实验方法或技术——Low field nuclear magnetic resonance (NMR) technique.

实验的材料——navel orange.

实验研究结果——A trend of T_2 variation was observed in the diagram T_2 vs. storage time. T_2 values arise in the initial stage and then changed little in the following steady period and fall quickly at the end before the samples decay during the period of storage. A T_2 declining quickly processing before decay were observed in nearly all samples.

结论与意义——This method would be helpful for evaluating the quality of navel orange and the effect of preserving agent during storage in the future.

(2) 指示性摘要 (Indicative abstract)

有时被称为描述性摘要 (Indicative abstract)。这类摘要旨在阐明文章的主题，使广大读者很容易地做出是否阅读全文的决定。然而，由于其描述性而不是其本质，因而很少是全文的替代物。因此，指示性摘要不应该写成研究论文的标题式摘要，但指示性摘要可以用于综述性论文、会议报告、政府报告文献等，这类指示性摘要对文献数据库通常具有较大的价值。

(3) 报道-指示性摘要 (Informative-Indicative abstract)

至于报道-指示性摘要是介于上述两者之间的一种摘要形式，主要是以报道性摘要的形式表述信息价值较高的内容，而以指示性摘要的形式表述其余内容。当前国际国内的科技论文主要采用报道性摘要。

(4) 结构式摘要 (Structured abstracts)

不同于其他传统摘要 (Traditional abstracts)，结构式摘要应包括文题、作者署名、作者单位、摘要正文和关键词，按层次列出项目名称，逐项编写，重要信息不能漏掉。摘要正文一般包括 Research Objective (研究目的)、Method (研究方法)、Results (结果) 和 Conclusions (结论) 等项内容，有时可根据需要进行增减或合并。这类摘要优点就是可以使审稿便捷，核对方便，读者阅读容易 (Readability)，很快找到所需要的文章。结构性摘要又可使计算机检索更为准确有效，将传统出版物与电子数据库很好地关联起来。

例如：

Surviving and thriving in academia: a selective bibliography for new faculty members

Deborah Lee

Reference Services Review

Vol. 31. No. 1

Purpose

To provide a selective bibliography for graduate students and new faculty members with sources which can help them develop their academiccareer

Design/methodology/approach

A range of recently published (1993—2002) works, which aim to provide practical advice rather than theoretical books on pedagogy or educational administration, are critiqued to aid the individual make the transition into academia…

Findings

To provide information about each source, indicating what can be found there and how the information can help. To recognize the lack of real training of many academics before they are expected to take on teaching/researching duties and find some texts that help.

Research limitations/implications

It is not an exhaustive list and apart from one UK book all the rest are US publications which perhaps limits its usefulness elsewhere.

Practical implications

A very useful source of information and impartial advice for graduate students planning to continue in academia or for those who have recentlyobtained a position in academia.

Originality /value

This paper fulfils an identified information/resources need and offers practical help to an individual starting out on an academic career

Keywords: Bibliography, Higher education, Teachers, Academic staff, Research, Publishing

11.3.2 摘要的撰写要求

(1) 简明扼要

英文摘要内容包括研究的目的、方法，结果和结论，它具有相对的独立性、自明性，能量化的尽量加以量化，给读者提供主要的信息，不要把与论文无关或关系不大，以及本学科已成为常识性的东西写在英文摘要里。

(2) 要求使用法定计量单位以及正确地书写规范字和标点符号

众所周知的国家、机构、专用术语尽可能用简称或缩写；为方便检索系统转录，应尽量避免使用图、表、化学结构式、数学表达式、角标和希腊文等特殊符号。

(3) 摘要的长度

ISO 规定，大多数实验研究性文章，字数在 1000～5000 字的，其摘要长度限于 100～250 个英文单词。

(4) 摘要的时态

摘要所采用的时态因情况而定，应力求表达自然、妥当。写作中可大致遵循以下原则：

① 介绍背景资料时，如果句子的内容是不受时间影响的普遍事实，应使用现在时；如果句子的内容为对某种研究趋势的概述，则使用现在完成时。

② 在叙述研究目的或主要研究活动时，如果采用“论文导向”，多使用现在时（如：This paper presents…）；如果采用“研究导向”，则使用过去时（如：This study investigated…）。

③ 概述实验程序、方法和主要结果时，通常用现在式，如：We describe a new molecular approach to analyzing…

④ 叙述结论或建议时，可使用现在式、臆测动词或 may，should，could 等助动词，如：We suggest that climate instability in the early part of the last interglacial may have…

(5) 摘要的人称和语态

作为一种可阅读和检索的独立使用的文体，摘要一般只用第三人称而不用其他人称来写。有的摘要出现了“我们”“作者”作为陈述的主语，这会减弱摘要表述的客观性，有时也会出现逻辑上讲不通。由于主动语态的表达更为准确，且更易阅读，因而目前大多数期刊都提倡使用主动态，国际知名科技期刊“Nature”“Cell”等尤其如此。

现通过一个食品科学英文摘要的具体实例，说明英文摘要的内容和写作方法。

Low field nuclear magnetic resonance (NMR) technique was used to evaluate the quality and behavior of water in navel orange during the course of storage. Carr-Purcell-Meiboom-Gill (CPMG) sequence was used to acquire the samples' NMR echo signals. The spin-spin relaxation time (T_2) of samples were calculated by a mult-exponential regress model during the deferent period of storage until decay. A trend of T_2 variation was observed in the diagram T_2 vs. storage time. T_2 values arise in the initial stage and then changed little in the following steady period and fall quickly at the end before the samples decay during the period of storage. A T_2 declining quickly processing before decay were observed in nearly all samples. This method would be helpful for evaluating the quality of navel orange and the effect of preserving agent during storage in the future.

此例回答了如下四个问题

做了什么?（研究工作范围）

——用低场核磁（NMR）研究脐橙储藏期品质。

怎样做的?（实验要点）

——无损检测 T_2 弛豫时间。

结果如何?（主要结果）

——脐橙果实 T_2 弛豫时间能有效反映果实品质变化。

原因何在?（结果的解释，结论）

——脐橙果实存在不同的水。

下面就回答上述四个问题列举一些基本句式。

① 研究工作范围

(to be) carried out, performed, made, conducted, studied, investigated, described, dealt with, elucidated, employed, prepared, synthesized, monitored, determined, measured, observed, examined, identified, characterized, proposed, calculated, evaluated, discussed.

② 实验要点

(to be) prepared by the reaction of…with…

Synthesized via reacting

Obtained via condensation

By means of, by using, over a wide range of…in the temperature range from…to…in the presence of…

③ 主要结果

(a) 用 that 从句（宾语从句）表示实验结果：

The results indicate that…

The results show that…

The results demonstrate that…

The results reveal that…

(b) 用 It 引起的 that 从句：

It was shown that…

It can be seemed that…

It was discovered that…

It was concluded that…

It was suggested that…

It was supposed that…

It has become apparent that…

(c) 表示具有某种影响以及数量的依赖、增减关系：

… (to be) correlated with…

Associated with…

Related to…

Dependent upon…

Proportional to…

In a linear relationship with…

(d) (不) 相符的表示方法：

(to be) in good agreement with…

(to be) Found to agree well with…

(to be) Consistent with…

(to be) Found to coincide essentially with…

Good agreement (to be) found between…

(e) 结果解释 (结论)：

(to be) due to

(to be) attributed to

(to be) assigned to…

(to be) interpreted on the basis of…

Can be caused by…

11.4 关键词

关键词 (Key Words) 是为了对文章进行检索而做的标引，即赋予某篇文献的检索标识。对关键词的正确选定有助于提高检索效率，即查全率、查准率，避免漏检、误检。关键词是从标题、摘要和内容中选取出来的揭示主题内容，并经优选和取舍而构成的关键性词汇。通常每篇论文选出 3～8 个词。

11.4.1 关键词的选择

从研究的对象、性质和采取的方法（手段）选取关键词。如题名为 Thermal Degradation of Some Polyimides（几种聚酰亚胺的热裂解），其关键词为：polyimides（聚酰亚胺，系指研究对象），thermal stability（热稳定性，指性质），thermogravimetry，pyrolysis，gas chromatography，mass spectrometry（热重法、裂解、气相色谱、质谱，均指研究

方法）。合理选用关键词会有利于该篇论文被检出和引用。通过关键词的逻辑组合可以清晰地提示论文的主要内容。

11.4.2 关键词的排序

有时，关键词主要来自文题，并以其在文题中出现的顺序排列，例如：

例1 文题：查尔酮冠醚PVC膜钾离子选择性电极的初步研究

关键词：查尔酮冠醚，PVC，膜，钾离子，电极

通常，关键词不应简单地按照在文章中出现的先后次序排序。例如：

例2 文题：均一沉淀法TiO_2被覆云母片

关键词：均一沉淀法，TiO_2，膜，覆云母片

此文报道的是在天然矿物云母片上被覆盖了TiO_2薄膜。关键词首先应提出研究的对象，然后是采用的方法。文题和关键词的排序修改为：

文题：均一沉淀法云母片被覆TiO_2

关键词：云母片，TiO_2，均一沉淀法

关键词的排序应符合人们的逻辑思维。

例如："聚氯乙烯-乙丙橡胶-氯丁橡胶三元共混物的研究"一文关键词的原顺序为"相容性，共混物，聚氯乙烯，乙丙橡胶，氯丁橡胶"。试想，无聚氯乙烯等3个组分哪来的共混物，而无共混物相容性又从何谈起？因此关键词的正确排序应为："聚氯乙烯，乙丙橡胶，氯丁橡胶，共混物，相容性"。

11.4.3 选取关键词的注意事项

某些学科（如化学）通常不选"……和……""……的……"乃至短语，可将以词组或短语形式出现的关键词进行拆分，检索时通过合理的逻辑组合，即通过计算机检索系统按特定的逻辑关系加以组配，可以获得所需要的信息。例如：动力学研究（kinetic study）。这里kinetic是形容词，修饰名词study，可简化为：动力学（kinetics）。热分解机理（thermal decomposition mechanism）修改为：热分解，机理（thermal decomposition，mechanism）。

关键词既应有一定的专指性，与文献的主题内容相符，有利于实现检索功能，又不至于因关键词的专指度过大而漏检。例如，含有关键词DSC的文章数是1000，含有water一词的文章数是3000，含有polysaccharides的文章数是500篇，将其结合DSC* water* polysaccharides则检索到的只有30篇。

不用非通用的代号和分子式作关键词，例如，下列各关键词应分别给出全称：CF复合材料应改为"碳纤维复合材料"，"ATP"应为"三磷酸腺苷"。而通用代号则可选作关键词，如DTA（差热分析）、IR（红外光谱）、NMR（核磁共振）等。

关键词需包含基本的主题内容如"CaO与SO_2反应的热重法研究"一文，原稿关键词为："热重分析，硫酸盐反应，石灰石"。本文的目的是为了消除燃煤锅炉排放的SO_2，采用的方法是：利用石灰石在锅炉中煅烧生成的CaO与烟气中的SO_2反应，将SO_2转化成$CaSO_4$，SO_2是本文论述的主题，在关键词中不应没有。因此，关键词及其排序应当是"SO_2，石灰石，硫酸盐化反应，热重法"。

避免过于空泛的词，如合成、测定、性质、方法、表征、分析、研究、探讨、关系之类的词过于空泛，缺乏特指，不利于检索，应避免选用。关键词应尽量选得具体些，如只是研究了一两种水果原料，如芒果、香蕉，则可直接将这两种水果选作关键词，而不是笼统地称为"水果"。

11.5 引言

引言（Introduction）位于正文的起始部分，主要叙述自己写作的目的或研究的宗旨，使读者了解和评估研究成果。主要内容包括：介绍相关研究的历史、现状、进展，说明自己对已有成果的看法，以往工作的不足之处，以及自己所做研究的创新性或重要价值；说明研究中要解决的问题、所采取的方法，必要时须说明采用某种方法的理由；介绍论文的主要结果和结构安排。

写作要求如下：

① 尽量准确、清楚且简洁地指出所探讨问题的本质和范围，对研究背景的阐述做到繁简适度。

② 在背景介绍和问题的提出中，应引用“最相关”的文献以指引读者。要优先选择引用的文献中包括相关研究的经典、重要和最具说服力的文献，力戒刻意回避引用的最重要的相关文献（甚至是对作者研究具某种“启示”性意义的文献），或者不恰当地大量引用作者本人的文献。

③ 采取适当的方式强调作者在本次研究中最重要的发现或贡献，让读者顺着逻辑的演进阅读论文。

④ 解释或定义专门术语或缩写词，以帮助编辑、审稿人和读者阅读稿件。

⑤ 常用的人称代词有第一人称（I，we）和第三人称（it，they），第二人称代词用的很少，避免使用隐晦的 the author，the authors，但是在文中不要使用过多的人称代词。适当地使用“I”“We”或“Our”，以明确地指示作者本人的工作，如：最好使用“We conducted this study to determine whether…”，而不使用“This study was conducted to determine whether…”。叙述前人工作的欠缺以强调自己研究的创新时，应慎重且留有余地。可采用类似如下的表达：To the author's knowledge…；There is little information available in literature about…；Until recently，there is some lack of knowledge about…，等等。

⑥ 时态运用：英语句子中作谓语的动词发生的时间和存在状态不同，其表达形式不同，谓语动词的这种不同的形式称作时态。英语句子中的谓语动词虽然有 16 个时态，但在科技论文中常用的只有 5 种，即一般现在时、一般过去时、一般将来时、过去完成时和现在完成时。

(a) 叙述有关现象或普遍事实时，句子的主要动词多使用现在时，如：“little is known about X”或“little literature is available on X”“streptomycin inhibits the growth of tuber culosis”

(b) 描述特定研究领域中最近的某种趋势，或者强调表示某些“最近”发生的事件对现在的影响时，常采用现在完成时，如：“few studies have been done on X”或“little attention has been devoted to X”。如：“though immobilization does not necessarily lead to stabilization，there have been many reports on enzyme stabilization by immobilization”

(c) 在阐述作者本人研究目的的句子中应有类似 This paper，The experiment reported here 等词，以表示所涉及的内容是作者的工作，而不是指其他学者过去的研究。

例如：“In summary，previous methods are all extremely inefficient. Hence a new approach is developed to process the data more efficiently.”就容易使读者产生误解，其中的第二句应修改为：“In this paper，a new approach will be developed to process the data more efficiently.”或者，“This paper will present (presents) a new approach that process the data more efficiently.”

11.6 材料和方法（Materials and Methods）

在论文中，这一部分用于说明实验的对象、条件、使用的材料、实验步骤或计算的过程、公式的推导、模型的建立等。对过程的描述要完整具体，符合其逻辑步骤，以便读者重复实验。

具体要求如下：

(1) 对材料的描述应清楚、准确　材料描述中应该清楚地指出研究对象（样品或产品、动物、植物、病人）的数量、来源和准备方法。对于实验材料的名称，应采用国际同行所熟悉的通用名，尽量避免使用只有作者所在国家的人所熟悉专门名称。

(2) 对方法的描述要详略得当、重点突出　应遵循的原则是给出足够的细节信息以便让同行能够重复实验，避免混入有关结果或发现方面的内容。如果方法新颖、且不曾发表过，应提供所有必需的细节；如果所采用的方法已经公开报道过，引用相关的文献即可（如果报道该方法期刊的影响力很有限，可稍加详细地描述）。

(3) 力求语法正确、描述准确　由于材料和方法部分通常需要描述很多的内容，因此通常需要采用很简洁的语言，故使用精确的英语描述材料和方法是十分重要的。需要注意的方面通常有：①不要遗漏动作的执行者，如：“To determine its respiratory quotient, the organism was…”，显然，the organism 不能来 determine；又如：“Having completed the study, the bacteria were of no further interest.”，显然，the bacteria 不会来 completed the study。②在简洁表达的同时要注意内容方面的逻辑性，如：“Blood samples were taken from 48 informed and consenting patients…the subjects ranged in age from 6 months to 22 years”，其中的语法没有错误，但 6 months 的婴儿能表达 informed consent？③如果有多种可供选择的方法能采用，在引用文献时提及一下具体的方法，如：“cells were broken by as previously described”不够清楚，应改为：“cells were broken by ultrasonic treatment as previously described”。

(4) 时态与语态的运用　①若描述的内容为不受时间影响的事实，采用一般现在时，如：A twin-lens reflex camera is actually a combination of two separate camera boxes。②若描述的内容为特定、过去的行为或事件，则采用过去式，如：The work was carried out on the Imperial College gas atomizer, which has been described in detail elsewhere。③方法章节的焦点在于描述实验中所进行的每个步骤以及所采用的材料，由于所涉及的行为与材料是讨论的焦点，而且读者已知道进行这些行为和采用这些材料的人就是作者自己，因而一般都习惯采用被动语态。例如，优：“The samples were immersed in an ultrasonic bath for 3 minutes in acetone followed by 10 minutes in distilled water.”；劣：“We immersed the samples in an ultrasonic bath for 3 minutes in acetone followed by 10 minutes in distilled water.”。④如果涉及表达作者的观点或看法，则应采用主动语态。例如，优：“For the second trial, the apparatus was covered by a sheet of plastic. We believed this modification would reduce the amount of scattering.”；优：“For the second trial, the apparatus was covered by a sheet of plastic to reduce the amount of scattering.”；劣：“For the second trial, the apparatus was covered by a sheet of plastic. It was believed that this modification would reduce the amount of scattering.”。

11.7 结果（Results）

本部分描述研究结果，它可自成体系，读者不必参考论文其他部分，也能了解作者的研究成果。对结果的叙述也要按照其逻辑顺序进行，使之既符合实验过程的逻辑顺序，又符合实验结果的推导过程。本部分还可以包括对实验结果的分类整理和对比分析等。

写作要求如下：

（1）对实验或观察结果的表达要高度概括和提炼，不能简单地将实验记录数据或观察事实堆积到论文中，尤其是要突出有科学意义和具代表性的数据，而不是没完没了地重复一般性数据。

（2）对实验结果的叙述要客观真实，即使得到的结果与实验不符，也不可略而不述，而且还应在讨论中加以说明和解释。

（3）数据表达可采用文字与图表相结合的形式。如果只有一个或很少的测定结果，在正文中用文字描述即可；如果数据较多，可采用图表形式来完整、详细地表述，文字部分则用来指出图表中资料的重要特性或趋势。切忌在文字中简单地重复图表中的数据，而忽略叙述其趋势、意义以及相关推论。

（4）适当解释原始数据，以帮助读者理解。尽管对于研究结果的详细讨论主要出现在“讨论”章节，但“结果”中应该提及必要的解释，以便让读者能清楚地了解作者此次研究结果的意义或重要性。

（5）文字表达应准确、简洁、清楚。避免使用冗长的词汇或句子来介绍或解释图表。为简洁、清楚起见，不要把图表的序号作为段落的主题句，应在句子中指出图表所揭示的结论，并把图表的序号放入括号中。例如，“Figure 1 shows the relationship between A and B”不如“A was Significantly higher than B at all time points hecked (Figure 1)”，又如，“It is clearly shown in Table 1 that nocillin inhibited the growth of *N. gonorrhoeae*.”不如“Nocillin inhibited the growth of *N. gonorrhoeae* (Table 1)”。

（6）时态的运用　①指出结果在哪些图表中列出，常用一般现在时，如：“Figure 2 shows the variation in the temperature of the samples over time.”。②叙述或总结研究结果的内容为关于过去的事实，所以通常采用过去时，如：“After flights of less than two hours, 11% of the army pilots and 33% of the civilian pilots reported back pain.”。③对研究结果进行说明或由其得出一般性推论时，多用现在时，如：“The higher incidence of back pain in civilian pilots may be due to their greater accumulated flying time.”。④不同结果之间或实验数据与理论模型之间进行比较时，多采一般现在时（这种比较关系多为不受时间影响的逻辑上的事实），如：“These results agree well with the findings of Smith, et al.”。

11.8 讨论（Discussion）

“讨论”的重点在于对研究结果的解释和推断，并说明作者的结果是否支持或反对某种观点、是否提出了新的问题或观点等。因此撰写讨论时要避免含蓄，尽量做到直接、明确，以便审稿人和读者了解论文为什么值得引起重视。讨论的内容主要有：①回顾研究的主要目的或假设，并探讨所得到的结果是否符合原来的期望？如果没有的话，为什么？②概述最重要的结果，并指出其是否能支持先前的假设以及是否与其他学者的结果相互一致；如果不是的话，为什么？③对结果提出说明、解释或猜测；根据这些结果，能得出何种结论或推论？④指出研究的限制以及这些限制对研究结果的影响；并建议进一步的研究题目或方向；⑤指出结果的理论意义（支持或反驳相关领域中现有的理论、对现有理论的修正）和实际应用。

具体的写作要求如下：

（1）对结果的解释要重点突出，简洁、清楚。为有效地回答研究问题，可适当简要地回顾研究目的并概括主要结果，但不能简单地罗列结果，因为这种结果的概括是为讨论服务的。

（2）推论要符合逻辑，避免实验数据不足以支持的观点和结论。根据结果进行推理时要适度，论证时一定要注意结论和推论的逻辑性。在探讨实验结果或观察事实的相互关系和科学意义时，无需得出试图去解释一切的巨大结论。如果把数据外推到一个更大的、不恰当的结论，不仅无益于提高作者的科学贡献，甚至现有数据所支持的结论也受到怀疑。

（3）观点或结论的表述要清楚、明确。尽可能清楚地指出作者的观点或结论，并解释其支持还是反对早先的工作。结束讨论时，避免使用诸如“Future studies are needed.”之类苍白无力的句子。

（4）对结果科学意义和实际应用效果的表达要实事求是，适当留有余地。避免使用“For the first time”等类似的优先权声明。在讨论中应选择适当的词汇来区分推测与事实。例如，可选用“prove”“demonstrate”等表示作者坚信观点的真实性；选用“show”“indicate”“found”等表示作者对问题的答案有某些不确定性；选用“imply”“suggest”等表示推测；或者选用情态动词“can”“will”“should”“probably”“may”“could”“possibly”等来表示论点的确定性程度。

（5）时态的运用　①回顾研究目的时，通常使用过去时。如：“In this study, the effects of two different learning methods were investigated.”。②如果作者认为所概述结果的有效性只是针对本次特定的研究，需用过去时；相反，如果具有普遍的意义，则用现在时。如：“In the first series of trials, the experimental values were all lower than the theoretical predictions. The experimental and theoretical values for the yields agree well.”。③阐述由结果得出的推论时，通常使用现在时。使用现在时的理由是作者得出的是具普遍有效的结论或推论（而不只是在讨论自己的研究结果），并且结果与结论或推论之间的逻辑关系为不受时间影响的事实。如：“The data reported here suggest (These findings support the hypothesis, our data provide evidence) that the reaction rate may be determined by the amount of oxygen available.”。

11.9 结论（Conclusions）

作者在文章的最后要单独用一章节对全文进行总结，其主要内容是对研究的主要发现和成果进行概括总结，让读者对全文的重点有一个深刻的印象。有的文章也在本部分提出当前研究的不足之处，对研究的前景和后续工作进行展望。应注意的是，撰写结论时不应涉及前文不曾指出的新事实，也不能在结论中重复论文中其他章节中的句子，或者叙述其他不重要或与自己研究没有密切联系的内容，以故意把结论拉长。

11.10 致谢（Acknowledgement）

出于对帮助者的礼貌和感谢，一般在论文结尾处要写上某些致谢用语。在研究及论文写作期间给予明显帮助和合作，或虽非直接参与但有直接影响的机构、团体与个人都属于书面致谢之列，一般置于结论之后、参考文献之前。其基本形式如下：

致谢者　被致谢者　原因

例如：J. Ma is very grateful to the National Science Foundation of China (NNSFC) for the support.

也可以是作者具体指出某人做了什么工作使研究工作得以完成，从而表示谢意。

如果作者既要感谢某机构、团体、企业或个人的经济资助，又要感谢他人的技术、设备的支持，则应按惯例先对经济资助表示感谢，再对技术、设备支持表示感谢。

致谢的文字表达要朴素、简洁，以显示其严肃和诚意。

常用的致谢句型有：

be grateful to (somebody) for…

be indebted to (somebody) for…

thank (somebody) for…

acknowledge (something or somebody) for…

appreciate (something) …

own gratitude to (somebody) for…

a special thank is due to (somebody) for…

would like to express one's appreciation to (somebody) for…

wish to thank (somebody) for…

11.11 参考文献 (References)

关于参考文献的内容和格式，建议作者在把握参考文献著录基本原则的前提下，参阅所投刊物的“投稿须知”中对参考文献的要求，或同一刊物的其他论文参考文献的注录格式，使自己论文的文献列举和标注方法与所投刊物相一致。这里只对基本规则作简单介绍。

参考文献的类型有专著、连续出版物、电子文献三类。

例如：Hatakeyama T, Quinn F X, Thermal Analysis: Fundamentals and Applications to Polymer Science [M], 2nd Ed. Chichester: John Wiley, 1999.

唐学明，陈真宝．双烯烃配位聚合进展 [M] //黄葆同，沈之荃，等．烯烃配位聚合进展．北京：科学出版社，1998：172-202.

ISO 5966—1982 中规定参考文献应包含以下三项内容：作者/题目/有关出版事项。其中出版事项包括：书刊名称、出版地点、出版单位、出版年份以及卷、期、页等。

参考文献的具体编排顺序有两种：

① 按作者姓氏字母顺序排列 (alphabetical list of references)；

② 按序号编排 (numbered list of references)，即对各参考文献按引用的顺序编排序号，正文中引用时只要写明序号即可，无需列出作者姓名和出版年代。

目前常用的正文和参考文献的标注格式有三种：

(1) MLA 参考文献格式　MLA 参考文献格式由美国现代语言协会 (Modern Language Association) 制定，适合人文科学类论文，其基本格式为：在正文标注参考文献作者的姓和页码，文末间列参考文献项，以 Works Cited 为标题。

(2) APA 参考文献格式　APA 参考文献格式由美国心理学会 (American Psycho-

logical Association）制定，多适用于社会科学和自然科学类论文，其基本格式为：正文引用部分注明参考文献作者姓氏和出版时间，文末单列参考文献项，以 References 为标题。

（3）Chicago参考文献格式　该格式由芝加哥大学出版社（University of Chicago Press）制定，可用于人文科学类和自然科学类论文，其基本格式为：正文中按引用先后顺序连续编排序号，在该页底以脚注（Footnotes）或在文末以尾注（Endnotes）形式注明出处，或在文末单列参考文献项，以 Bibliography 为标题。

下面举 ISO 推荐的哈佛标引格式为例：

① 期刊文章

Dicken CH，Connodly SM. 1993. A randomized crossover comparison. son of two low-dose contraceptives：effects on serum lipids and lipoprotein. Nature（London）371：453-456.

作者名．出版年份．文章名．刊物名称（出版地）及期卷号：参考页码

② 书籍或书中章节

Burry KA. 1994. Urban economics and public policy. New York：St Matin's Press. pp 380.

作者名. 出版年份. 书籍名. 出版地：出版社名. 参考页码

③ 非英语文章　一般非英语文献要把文题译成英文。也有将原文标题列出，但译文跟在后面括号内。

Wang KL. 1992. Molecules and ife：an introduction to molecules biology. Beijing：Science press. p78-84. Translated from Chinese by Li M.

注意事项：

① 文献的时限性。作者应引用亲自阅读过的那些较为重要的最新文献，多引近5年内发表的最新文献，如达50%～70%，这至少可以说明该研究课题能很好跟踪当前本学科的发展。

② 避免直接在标题上标注。多出现在根据外文版（或中文版）的专著编写的中文书籍的有关章节。例如："5.4.1 线团状分子[40]，5.4.3 蠕虫状链[40,44]"。而后在文献表中分别给出了文献［40］和［44］。应在正文中对引文作者或引语标注序号。

③ 切记有意漏引。为了避免自我炫耀甚至剽窃之嫌，应客观真实地反映科学的继承与发展。

④ 参考文献的数量。据悉，几年来，我国科技期刊论文的篇均参考文献数量呈上升趋势，但仍低于国外科技期刊论文的篇均参考文献数。根据期刊引证报告（JCR）2000 版报道的数据，国际期刊、中国期刊各45种的篇均引文数分别为28条和14条。

11.12 其他

11.12.1 附录（Appendix）

主要是冗长定理的证明，以及实验中装置的冗长描述及参数等。

11.12.2 注释（Notes）

注释用于补充说明正文中某些需解释但不适合在正文中叙述的内容。注释可以为脚注或尾注形式，其内容可包括相关背景、人物、专有名称的解释，也可作为参考文献的一种列写

形式。当以后者形式出现时，其书写形式遵循参考文献注录的基本格式，只是每一条注释都应加有编号。

11.12.3 符号和术语 (Notation/Nomenclature)

有些刊物要求文章作者把本文中出现的各种符号、希腊字母所代表的含义单独列出，并标明为 Notation 或 Nomenclature，以便读者参阅。该部分一般放在正文之后、参考文献之前，也有的放在引言之后，甚至可能不出现 Notation 或 Nomenclature 字样，只用一方框列出，而通用符号可以不作解释。

References

[1] Carol A Wallace，William H Sperber，Sara E Mortimore. Food safety for the 21 st century：managing HACCP and food safety throughout the global supply chain. Wiley-blackwell，2011.

[2] Parker R. 食品科学导论. 影印版. 北京：中国轻工业出版社，2005.

[3] 张兰威，李佳新. 食品科学与工程英语. 哈尔滨：哈尔滨工程大学出版社，2005.

[4] Food Safety and Nutrition Resources for Healthcare Professionals http：//www. fda. gov/food/resourcesforyou/healthcare-professionals/default. htm.

[5] U. S. Food and Drug Administration. NUTRITION. http：//www. fda. gov/Food/DietarySupplements/default. htm.

[6] USDA. Food and Nutrtion https：//www. usda. gov/wps/portal/usda/usdahome? navid=food-nutrition.

[7] Overview of Food Ingredients，Additives & Colors. www. fda. gov/Food/IngredientsPackagingLabeling/FoodAdditivesIngredients/ucm094211. htm＃qalabel.

[8] Food and Drug Administration. www. fda. gov.

[9] U. S. Department of Agriculture Food Safety and Inspection Service. www. fsis. usda. gov.

[10] Food Additives Information. https：//www. fsis. usda. gov/wps/portal/redirection? url=/Fact _ Sheets/Additives.

[11] 郝利平，聂乾忠，周爱梅. 食品添加剂. 第3版. 北京：中国农业大学出版社，2016.

[12] Tommy L Wheeler，Linda S Papadopoulos，Rhonda K Miller，et al. Research guidelines for cookery，sensory evaluation，and instrumental tenderness measurements of meat，Second Edition Version 1. 02. American Meat Science Association，2016.

[13] AMSA (1978). Guidelines for cookery and sensory evaluation of meat. Chicago，IL：American Meat Science Association. 1978.

[14] AMSA (1995). Research guidelines for cookery，sensory evaluation and instrumental tenderness measurements of fresh meat. Chicago，IL：American Meat Science Association. 1995.

[15] NAMI (2015). Meat buyers guide. Washington，DC. North American Meat Institute. 2015.

[16] Hiner R L，Madsen L L，Hankins O G. Histological characteristics，tenderness and drip losses of beef in relation to temperature of freezing. Journal of Food Science，1945 (10)，312-324.

[17] ASTM E1871 (2010). Standard guide for serving protocol for sensory evaluation of foods and beverages. West Conshohocken，PA：ASTM International. 2010.

[18] Bohnenkamp J S，Berry B W. Effect of sample numbers on sensory activity of a trained ground beef texture panel. Journal of Sensory Studies，1987 (2)：23.

[19] NIH (1979). The Belmont report. Ethical principles and guidelines for the protection of human subjects of research. Retrieved from http：//ohsr. od. nih. gov/guidelines/belmont. html.

[20] 黄吉武，董建. 毒理学基础. 第2版. 北京：人民卫生出版社，2016.

[21] Hsu CH，Stedeford T. Cancer risk assessment：chemical carcinogenesis，Hazard Evaluation，and Risk Quantification. N J Wiley：Hoboken，2010.

[22] Ahvenainen R. Gas packaging of chilled meat products and ready-to-eat food，Technical Research Centre of Finland，VTT，Espoo，Finland，1989.

[23] Ahvenainen R. Novel food packaging techniques：Elsevier 3. 2003；Begley T H，Hollifield，H C. Recycled polymers in food packaging：Migration considerations. Food Technology，1993，47 (11)：109-112.

[24] Brody A L. Modified Atmosphere Packaging，Institute of Packaging Professionals. USA：Virginia，1994.

[25] Bureau G，Multon J L. Food Packaging Technology：Volume Ⅰ and Ⅱ. USA：Wiley-VCH Inc，1995.

[26] Coles R，McDowell D，Kirwan M J. Food packaging technology：Vol 5. CRC Press，2003.

[27] Downes T W，Arndt G，Goff J W，et al. Factors affecting Seal Integrity of Aseptic Paperboard/Foil Packages. Aseptipak' 85：Proceedings of the Third International Conference and Exhibition on Aseptic Packaging，1985：363-401.

[28] Farber J M，Dodds K L. Principles of modified-atmosphere and sous vide product packaging. Lancaster-Basel：Technomic Publishing Co Inc，1995.

[29] Ferrante MA，Martin K. Safety and convenience take center stage. Food Engineer，1999，71 (11)：79-86.

[30] Ferrante MA. The brave new world of packaging technology at Pack Expo 96. Food Engineer，1996. 68 (10)：99-102.

[31] Han J H. Innovations in food packaging. Academic Press，2005.

[32] Higgins KT，Active packaging gets a boost. Food Engineer，2001，73 (10)：20.

[33] Hotchkiss J H，Tamper evident packaging for foods：Current technology. Proceedings Prepared Foods，1983：152 (10)：66-67.

[34] Klungness J H，Lin C H，Rowlands，R E，Contaminant removal from recycled wastepaper pulps. Pulping Conference proceedings，1990：1. 8-12.

[35] Martin K. Safety and convenience take center stage. Food Engineer，1999，71 (11)：79-86.

[36] Piergiovanni Luciano，Sara Limbo. Food Packaging Materials. Springer Verlag，2016.

[37] Sloan AE. The silent salesman. Food Technology，1996，50 (12)：25.

[38] Plant and grounds，食品生产加工企业的厂房与地面. 食品伙伴网 http：//www. foodmate. net/.

[39] Food safety modernization ac. 食品安全网，http：//www. foodsafe. net/article. asp? nameid=181.

[40] Code of Federal Regulations (CFR) http：//www. foodmate. net/.

[41] U. S. Food and Drug Administration. http：//www. cfsan. fda. gov/～lrd/haccp. html.

[42] 食品生产通用卫生规范 GB 14881—2013. http：//www. lawtime. cn/info/shipin/fagui/201311282873376. html.

[43] INTRODUCTION. List of NACE codes (2010). http：//ec. europa. eu/competition/mergers/cases/index/nace _ all. html.

[44] Bricout A，Carlier A，Lanau A，Harison R，Hui S. Food Safety Assurance Systerm. Les spécificités de la logistique alimentaire，Université de Picardie Jules Verne，Institut National Supérieur des Sciences et Techniques de Saint-Quentin，France，2009.

[45] Principles and Systems for Quality and Food Safety Management. Chapter 22-Principles and Systems for Quality and Food Safety Management. A Practical Guide for the Food Industry，2014：537-558.

[46] 5S management system. http：//leanmanufacturingtools. org/192/what-is-5s-seiri-seiton-seiso-seiketsu-shitsuke/.

[47] GLOBALG. A. P. http：//www. globalgap. org/uk _ en/what-we-do/globalg. a. p. -certification/.

[48] GAP. Challenges of Developing Countries in Complying Quality and Enhancing Standards in Food Industries，Procedia-Social and Behavioral Sciences：224，2016：445-451.

[49] HACCP. http：//www. omafra. gov. on. ca/english/food/foodsafety/processors/haccp. htm.

[50] GMP. http：//www. wiley. com/WileyCDA/WileyTitle/productCd-111831820X. html.

[51] ISO，http：//www. iso. org/iso/home. html.

[52] ISO 22000，http：//www. iso. org/iso/home/store/catalogue _ tc/catalogue _ detail. htm? csnumber=35466.

[53] CNCLUSION. http：//www. sciencedirect. com/science/book/9780123815040.

[54] CAC. http：//www. fao. org/fao-who-codexalimentarius/en/.

[55] CAC. http：//www. fao. org/fao-who-codexalimentarius/codex-home/zh/.

[56] Kotwal V. Codex Alimentarius Commission：Role in International Food Standards Setting. Encyclopedia of Food and Health，2016 (6)：197-205.

[57] Heilandt T，Mulholland C A，Younes M. Institutions Involved in Food Safety：FAO/WHO Codex Alimentarius Commission (CAC). Encyclopedia of Food Safety，2014 (4)：345-353.

[58] Codex Alimentarius Commission Procedural Manual，22nd ed. Rome：WHO and FAO of the United Nations，2014.

[59] Enhancing participation in Codex activities，an FAO/WHO training package. FAO and WHO. 2005.

[60] Codex Alimentarius Commission http：//www. codexalimentarius. org/codex-home/en/. FAO/WHO Codex Training Manual and Codex E-Learning Course.

[61] Joint FAO/WHO Food Standards Programme. Codex Alimentarius Commission Procedural Manual 20th edn. Rome：Food and Agriculture Organization of the United Nations and World Health Organization，2001.

[62] Joint FAO/WHO Food Standards Programme. Understanding the Codex Alimentarius，3rd edn. Rome：Food and Agriculture Organization of the United Nations and World Health Organization，2011.

[63] Masson-Matthee MD. The CodexAlimentariusCommissionandits Standards. The Hague：T M C Asser Press，2007.

[64] Jorge Barros-Velázquez. ELSEVIER，2016. Antimicrobial Food Packaging.

[65] GMPs - Section One：Current Food Good Manufacturing Practices http：//www. fda. gov/Food/GuidanceRegula-

tion/CGMP/ucm110907. htm.

[66] Current Good Manufacturing Practice in Manufacturing, Packing, or Holding Human Food http: //www. accessdata. fda. gov/SCRIPTs/cdrh/cfdocs/cfcfr/CFRSearch. cfm? CFRPart=110.

[67] Eight key conditions of a sanitation SOP. http: //msue. anr. msu. edu/news/eight _ key _ conditions _ of _ a _ sanitation _ sop.

[68] Sanitation standard operating procedures (21CFR120. 6). http: //www. accessdata. fda. gov/scripts/cdrh/cfdocs/cfcfr/CFRSearch. cfm? fr=120. 6.

[69] Hazard Analysis and Critical Control Point (HACCP) System and Guidelines for Its Application. http: //www. fao. org/docrep/005/y1579e/y1579e03. htm#bm3.

[70] ISO 9000 quality management. http: //www. iso. org/iso/home/standards/management-standards/iso _ 9000. htm.

[71] ISO 22000 food safety management. http: //www. iso. org/iso/home/standards/management-standards/iso22000. htm.

[72] 陈宗道，刘金福，陈绍军. 食品质量与安全管理. 第2版. 北京：中国农业大学出版社，2011.

[73] 刘振海，刘永新，陈忠才，等. 中英文科技论文写作. 北京：高等教育出版社，2012.

[74] 刘振聪，修月祯. 英语学术论文写作. 第2版. 北京：中国人民大学出版社，2013.

[75] 金国斌，李蓓蓓. 新编包装科技英语. 北京：中国轻工业出版社，2013.

[76] 吴江梅，黄佩娟，马平，武田田，刘真. 英语科技论文写作. 北京：中国人民大学出版社，2013.

[77] http: //blog. sina. com. cn/s/blog _ 699f2bd701010doz. html.

[78] 周序林，何均洪. 基于简明原则的科技论文英文摘要写作规范. 西南民族大学学报，2016，42 (4)：469-470.